Towards an Environment Research Agenda

Also edited by Alyson Warhurst

TOWARDS A COLLABORATIVE ENVIRONMENT RESEARCH AGENDA

Towards an Environment Research Agenda

A Second Selection of Papers

Edited by

Adrian Winnett
International Centre for the Environment
University of Bath

and

Alyson Warhurst
Corporate Citizenship Unit
University of Warwick

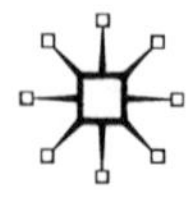

First published 2003 by
PALGRAVE MACMILLAN
Houndmills, Basingstoke, Hampshire RG21 6XS and
175 Fifth Avenue, New York, N.Y. 10010
Companies and representatives throughout the world

PALGRAVE MACMILLAN is the global academic imprint of the Palgrave Macmillan division of St. Martin's Press, LLC and of Palgrave Macmillan Ltd. Macmillan® is a registered trademark in the United States, United Kingdom and other countries. Palgrave is a registered trademark in the European Union and other countries.

ISBN 0–333–67480–4

This book is printed on paper suitable for recycling and made from fully managed and sustained forest sources.

A catalogue record for this book is available from the British Library.

Library of Congress Cataloging-in-Publication Data
Towards an environment research agenda: a second selection of papers/ edited by Adrian Winnett
p. cm.
Includes bibliographical references and index.
ISBN 0–333–67480–4 (cloth)
1. Environmental policy. 2. Pollution. 3. Environmental management.
I. Winnett, Adrian.
GE170 .T68 2002
363.7—dc21 2002075989

10 9 8 7 6 5 4 3 2 1
12 11 10 09 08 07 06 05 04 03

Printed and bound in Great Britain by
Antony Rowe Ltd, Chippenham and Eastbourne

Contents

Part IV Valuing the Environment

Part V Managing the Environment

Acknowledgements

We are very grateful to the ICE support staff, Carolina Salter and Yvette Haine, for respectively organizing the seminar series and editing the papers with unfailing efficiency and enthusiasm.

ADRIAN WINNETT

The editor, contributors and publishers are grateful to the *Canadian Journal of Environmental Education* for allowing reproduction of Chapter 4. The illustration on page 91 is copyright the artist, Clifford Harper.

Every effort has been made to contact all the copyright-holders, but if any have been inadvertently omitted the publishers will be pleased to make the necessary arrangement at the earliest opportunity.

Notes on the Contributors

Ingolfur Blühdorn is Senior Lecturer in Politics in the Department of European Studies, University of Bath. His research is at the interface of (eco-) political theory and environmental sociology. He has published widely on Green Parties, social movements and issues of ecological modernization. His most recent work explores the political potential of the eco-movement and current developments in protest politics in a wider sense.

Mark R.C. Doughty took his first degree in biological sciences at Pembroke College, Oxford University, before taking a Masters at Bath University. His thesis was entitled 'Use of the Ecological Footprint Theory as a Sustainability Indicator' and used Bath as case study for examining the application of Ecological Footprint thinking. Mark has been a professional ecologist and environmental consultant for over 5 years and has worked on numerous Environmental Impact Assessments and other environmental appraisal work. His current position as Biodiversity and Heritage Advisor with Thames Water involves provision of technical and legal input to the company's extensive capital expenditure programme as well as assisting in the development of their policy with regards to sustainable water resources, ISO14001 and other environmental policy. He is also responsible for implementing a large wetland habitat enhancement scheme. As a freelance ecologist he has also been involved in several projects abroad including CSIRO research into Forest Fire ecology in Queensland, Australia and a joint University of Oxford/University of Mexico City project investigating the role of canine carnivores in prairie dog colonies of Northern Mexico.

Steve Gough is a Lecturer in the Centre for Research in Education and the Environment at the University of Bath. His interest in the environmental impacts of tourism development, and the possible role of education in mitigating them, developed during nine years he spent living, working and researching in Borneo. Originally trained as an economist, he has participated in a number of international geographical expeditions in Papua New Guinea, Borneo and northern Norway. He is an educational consultant to the World Wildlife Fund (US).

Geoffrey P. Hammond is Professor of Mechanical Engineering and Head of the Aero-Thermodynamics Group at the University of Bath. He is a mechanical engineer with a multidisciplinary background, including environmental engineering and management. During the 1960s and early 1970s he worked as a design and development engineer in the UK refrigeration industry, before commencing an academic career at Uganda Technical College (under the auspices of Voluntary Services Overseas) teaching mainly in the field of applied thermodynamics. He held various academic appointments within the Applied Energy Group at Cranfield University (1976–89) before moving to the University of Bath, where he took up a new Professorship partially supported by British Gas plc. Geoffrey Hammond's research interests concern 'energy systems and environmental sustainability', and 'thermofluids and heat transfer'. He is the author or co-author of many research papers, and was the joint recipient of the Dufton silver medal for one of these publications. Professor Hammond has been Chair and/or a member of many Research Council committees covering the built environment, computational modelling, and postgraduate training. He sits on the editorial boards of three archival journals that publish material in the area of energy and the environment. In recent years, he has undertaken international consultancy assignments for government ministries and industrial R&D organizations in Sri Lanka and Taiwan on sustainable energy technologies and energy efficiency in the transport sector respectively. Outside the University Professor Hammond is a Patron and was a Founder Trustee (1995–8) of the Bath Environment Centre Limited [renamed 'envolve: partnerships for sustainability'], as well as being involved in several related partnership bodies in Bath and North-East Somerset (B&NES) concerned with environmental sustainability. In September 1998 he became a member of the Environment Agency's North Wessex Area Environment Group (Deputy Chairman, October 2000), which advises the Agency on the range of its regulatory activities. He was also the independent Chairman (2000–2) of the Combe Down Stone Mines Community Association (a company limited by guarantee), which has the task of building a partnership with the B&NES Council aimed at stabilizing hazardous eighteenth and nineteenth century mine workings on the outskirts of Bath.

Bradley S. Jorgensen is a Lecturer in Applied Social Psychology at the University of Bath, UK. His research interests lie in human–environment relationships, procedural justice in environmental management, the economic valuation of public goods, environmentally sustainable consumption, and corporate environmental responsibility and governance. Dr Jorgensen holds a BSc with First Class Honours and a PhD from Curtin University of Technology in Western Australia. His current research focus is on perceptions of fairness in environmental decision-making.

Eamonn Molloy is a Senior Research fellow at the Said Business School, University of Oxford, UK. He holds a PhD in Science and Technology Studies from the University of Lancaster and his research interests include technology, structure and reorganization.

Alan D.M. Rayner is an ecologist and thinker interested in understanding pattern, process and relationship in living systems. An accomplished naturalist best known for his scientific studies of fungi, he was President of the British Mycological Society in 1998. He has published six scientific books, including *Degrees of Freedom – Living in Dynamic Boundaries*, and over 130 scientific articles. A founder of Bath Bio*Art, he also produces and exhibits colourful oil paintings that reflect his scientific knowledge and sense of rapport with the natural world.

William Scott researches issues concerning learning in relation to sustainability: philosophical issues to do with the nature of education; practical (pedagogical and evaluative) issues about the process of learning/teaching; and ethical questions about the focus and limits of such educational interventions. He is a Fellow of the Royal Society of Arts, a trustee of the Living Earth Foundation and a member of the executive committee of the National Association of Environmental Education, and he also edits the academic journals *Environmental Education Research* and *Assessment and Evaluation in Higher Education*.

Andrew Stables is a Senior Lecturer within the Department of Education at the University of Bath and has written extensively on the relationships of language and environment in education. He worked as an English teacher in schools before taking up a post in the Education Department of the University of Wales Swansea in 1989. He moved to Bath in 1994.

Caedmon Staddon is Lecturer in the School of Geography and Environmental Management at the University of the West of England on issues related to Eastern Europe and natural resource management. He completed his PhD in 1996 at the University of Kentucky with a dissertation entitled 'Democratisation, Environmental Management and the Production of New Political Geographies in Bulgaria: a case study of the 1994–5 Sofia Water Crisis'. His current research focuses on the implications of forestry sector restructuring for community sustainable development.

A.R.D. Taylor is International Science Co-ordinator at Wetlands International, based in Wageningen, the Netherlands, and a member of the Ramsar Convention on Wetlands Scientific and Technical Review Panel. He was trained as a biologist, with a DPhil in plant biochemistry, and has worked as a university lecturer in East Africa, wetlands advisor to the Ugandan Government, and as a wetlands officer at Somerset County Council, UK. His main interests are in using strategic inventory, monitoring and assessment tools to enable wise use of wetlands and in the application of river basin stakeholder-led wetlands management.

Alyson Warhurst was the founding Director of the International Centre for the Environment at the University of Bath. She is now with the Corporate Citizenship Unit of the University of Warwick.

Adrian Winnett is Co-Director of the International Centre for the Environment. He is an economist with particular interests in natural resources and economic growth. He holds degrees from the London School of Economics and the University of East Anglia. Major projects in which he has been involved include the management of fisheries in South and South-East Asia, ethical investment in the UK, and waste management in Slovakia. He is undertaking a major research project for the European Commission on indicators for the new information economy.

Geof Wood is Head of Department of Economics and International Development, in addition to his role as the founder-director, of the Institute for International Policy Analysis (IFIPA). He is a sociologist, specializing in international development, with a regional focus on South Asia. His research over 34 years since graduation has included: development administration (Zambia); rural development (villages in Bihar, India and Bangladesh, and more recently northern Pakistan); irrigation systems and their common property management; rural natural resource management (fisheries, forests); microfinance; social mobilization and social development; governance; poverty and livelihoods in both urban and rural contexts; and the application of concepts from social policy to thinking about poverty eradication, welfare and development in poor countries, leading in particular to writing about rights and security. He has taught undergraduate and postgraduate courses in: sociology of developing societies; political sociology; agrarian change; sociology and anthropology of development (to Masters); livelihoods analysis (Masters); natural resource management and sustainable development (Masters). He founded the cross-university MA programme in Environmental Science, Policy and Planning.

Introduction

Adrian Winnett

This is a second selection of papers from the seminar series organized by the University of Bath's International Centre for the Environment (ICE). The first volume appeared under the editorship of the then Director, Professor Alyson Warhurst (Warhurst, 2000). Alyson subsequently moved to the University of Warwick's Corporate Citizenship Unit, and the Directorship of ICE was transferred to Professor Anil Markandya and Dr Adrian Winnett of the Department of Economics and International Development.

It is very largely due to Alyson's energy and enthusiasm that ICE was so successfully established and developed. In view of this, and since many of the papers in this second volume were first given during her directorship, we are pleased that she has agreed to act as joint editor.

The mission and objectives of ICE remain as defined by Alyson in her introduction to the first volume. ICE co-ordinates and publicizes interdisciplinary environmental research and education across the University of Bath; nearly all departments are involved. The major activities are to formulate research proposals for submission to grant-awarding bodies, and to develop seminars and courses on environmental issues.

Within this broad framework, as personnel and the university's agenda have changed, there have been some shifts in emphasis, especially towards environmental aspects of the built infrastructure and of the spatial economy of the South-West of England as compared with other regions in the European Union. One of the particular drivers of these shifts has been the development of a new campus in the fast-growing town of Swindon, in which ICE is playing a leading role. However, the activities of ICE are by no means restricted to these issues, and it encompasses a very broad range of activities.

For the most part, ICE operates as a 'virtual' centre. An exception is the seminar series 'Towards an Environmental Research Agenda' (TERA), which brings together both presenters and audiences from inside and outside the university. The present volume is representative of the approach of these seminars, again as previously explained by Alyson in her introduction to the first volume which is to locate disciplinary excellence within an interdisciplinary context.

Several of the papers in the present volume are by authors represented in the earlier volume and represent further evolution of their thinking. Others show them moving in new directions. But many of the papers are by new authors, often responding to the challenges created by the ICE approach.

Part I contains two highly original conceptual papers which explore fundamental issues in our understanding of the environment: Blüdhorn's paper is in part a response to Rayner's. This debate continues, and in the next volume there will be further dialogue involving new contributors.

In Part II, Stables' paper provides a bridge from these high-level conceptual debates into the educational agenda. Gough and Scott link the burgeoning field of ecotourism to educational initiatives.

Part III is concerned with the spatial dimensions of the environment. Hammond and Doughty explore urbanism through environmental footprints; Winnett outlines an agenda for understanding the impacts of globalization on resource-based industries.

Part IV takes a constructively critical look at two of the major techniques used by environmental analysts: life-cycle assessment (Molloy) and contingent valuation (Jorgensen).

Part V illustrates evolving problems of environmental management in three diverse situations, each characterized by complex histories of conflict and accommodation among stakeholders: the Somerset Levels (Taylor), Bulgaria (Staddon), and the Hindukush (Wood).

Reference

Warhurst, A. (2000) *Towards a Collaborative Environment Research Agenda*. Basingstoke: Palgrave Macmillan.

Part I

Environmental Philosophy and Politics

1
Inclusionality – An Immersive Philosophy of Environmental Relationships

Alan D.M. Rayner

Summary

For centuries, our understanding of how we relate to our environment has been impeded by the deliberate exclusion of context which comes from rationalistic modes of enquiry that place unrealistically discrete boundaries between 'insides' and 'outsides', 'subjects' and 'objects' and 'self' and 'other'. Now that the global impact of human technology has reached unprecedented scales, there is an urgent need to appreciate the implications of this exclusion and to develop a philosophical framework that enables us to attune more empathically with our living space. The participatory philosophy of 'inclusionality', in which all things are viewed as dynamic contextual inclusions, may help by enabling us to value the explicit focus of rational inquiry not as 'all there is', but rather as a powerful, high resolution tool. This tool, when complemented by the collective imagination and insights arising from many, diverse perspectives, can help clarify implicit, holographic reality.

Introduction

For all our technological and medical advances, we human beings continue to live uneasily with one another, other life forms and our surroundings. The symptoms of this human dis-ease are all too obvious. Huge disparities remain, and may even be increasing, between rich and poor, well nourished and malnourished. Huge tracts of natural environment are subject to destructive exploitation, pollution and alteration. We unwittingly bring about disease epidemics that spread on unprecedented scales through populations of people and domestic and wild animals and plants. Confrontational political and electoral systems persist in which the viewpoints of a large proportion – sometimes the majority – of the population are deliberately disregarded, resulting in widespread disempowerment and apathy. Big and small wars continue to rage. Stress, anxiety and depression afflict increasing numbers of people. New technologies such as genetic manipulation and cloning pose unanswerable questions that are overlooked in the rush for commercial gain.

Conflict abounds, both within and between our psyches. And, since conflict results from a breakdown of communication and consequent loss of relationship, the inference is that something is getting in the way, estranging us from our neighbours and surroundings. What is this something? What are its implications for the way that we approach our relationship with our environment? Can we, should we, change our approach? These are the intellectual challenges addressed in this chapter.

The problem of rational exclusivity and the need for an 'inclusional' view of environmental relationships

Increasingly, I have come to believe that this something, this barrier just referred to, is an *attitude problem*, which arises from the way we are perceptually and cognitively predisposed to view our environment, and reinforce this view in our philosophical and scientific paradigms and methodologies. The problem resides in our very own, much vaunted *rationality* – our focus within fixed frames of reference on discrete, explicit things, regardless of their context, and consequent separation of subjects from objects, insides from outsides and self from other. In particular, we have tended to view 'the' *environment* literally as our *surroundings*, something *outside of ourselves* that can somehow be separated off objectively for independent consideration so that we can *choose* whether and how much we should take it into account when planning our future actions. We set it *in competition* with social and economic interests, and more often than not put it low on our agenda of priorities, giving more weight to what we think are more 'immediate' concerns.

I unconsciously represented my feelings about this attitude problem almost 30 years ago, in two paintings (Figures 1.1 and 1.2).

Our rationalistic outlook causes us to exclude from consideration all outside our immediate focus, and so to *ignore* context, leaving us out of touch and undernourished in an intellectual and emotional desert of our own making. We regard life and the universe like a box of Lego blocks that can be sorted, assembled and disassembled: a fixed Euclidean reference frame of empty Cartesian space and absolute time in which independent objects collide, compete and stick together, but can't truly relate.

Figure 1.1 'Tropical Involvement' (oil painting on board, by Alan Rayner, 1972). This painting depicts the dynamic complexity of living systems. A turbulent river rushes between rock-lined banks from fiery, tiger-striped sunset towards unexpected tranquillity where it allows a daffodil to emerge from its shallows. A night-bird follows the stream past intricately interwoven forest towards darkness. A dragonfly luxuriates below a fruit-laden tree, bereft of leaves. Life is wild, wet and full of surprises.

It all seems so alluringly simple and logical – the only uncertainties lie in the randomness of independent events, but we think that statistics and risk analyses can help us to account for those. Moreover, this alluring simplicity fits in extremely well with our predatory and discriminatory predisposition to single things out from their context. Analytical left brain hemisphere at the ready, eyes facing forward on the front of our faces, giving us binocular vision and depth of field but little or no view to side or rear, we are great *sorter-outers*. And that is how we're prone to think the world works – by sorting things out – viewing like some self-centred voyeuristic outsider through a window pane and making discriminatory choices, but without ourselves being involved or included in the picture. Herein lies our devotion to quantification,

Figure 1.2 'Arid Confrontation' (oil painting on board by Alan Rayner, 1973). This painting depicts the limitations of unempathic, analytical methodology. At the end of a long pilgrimage, access to life is barred from the objective stare by the rigidity of artificial boundaries. A sun composed of semicircle and triangles is caught between Euclidean straight lines and weeps sundrops into a canalized watercourse. Moonlight, transformed into penetrating shafts of fear encroaches across the night sky above a plain of desolation. Life is withdrawn behind closed doors.

embedded in the discreteness of our number system and units of measurement as well as in *seemingly* great ideas like natural selection and genetic determinism.

However, the simplicity is an illusion, because, as is obvious to everybody, but which many prefer to turn a blind eye to, in reality nothing occurs in complete isolation. The discrete boundaries assumed or imposed by rational inquiry to keep things 'pure' and 'simple', free from contaminating subjectivity and environmental noise, are artefacts. And these artefacts may actually complicate and ration our understanding by starving us of what we need to know. Real boundaries are dynamic interfaces, places of opportunity for reciprocal transformation between intercommunicating insides and outsides over nested scales from subatomic to universal. They are not fully discrete limits. Features arise dynamically, through the inductive coupling of explicit contents with their larger implicit context, which, like a hologram, can only be seen partially and in unique aspect from any one fixed viewpoint.

By the same token, we humans are as immersed in and inseparable from our environmental context as a whirlpool in a water flow: our every explicit action implicitly depends upon and reciprocally induces transformation of this context. When I outfold my fingers, powered by energy I have assimilated from outside my body, the surrounding space reciprocally infolds. Far from being separate, outside of ourselves, our environment becomes us as we become it. It is truly our *living space* – as much our inheritance as our genes. By taking self-centred action, regardless of context, we put that inheritance at risk and ultimately conflict with our true Selves, driven on by the rationalism that continues to underpin much purely analytical science and legalistic thinking.

So, what can we do about this attitude problem, and the resultant conceptual trap that holds us in thrall? The first thing is to admit that it really is a problem, and it really is a trap. Then we need to know just how big a problem it is, and how unforgiving a trap. Then we have to find an imaginative way out of the trap.

I suggest that the problem is biggest when we are trying to view the larger picture and longer-term consequences of our environmental relationships. For then the crucial uncertainty that we face is not *randomness*, but *implications* – how the future will unfold as neglected outside influences come to bear and one thing induces another. And part of the trap lies in the fact that these influences *seem negligible* from the perspective of the smaller picture and short term. This encourages us to carry on regardless, thinking we can build from the small to the large, from the short to the long, which is impossible when the small and short has already excluded vital contextual information.

The potential enormity of the environmental damage caused by small, short-term thinking is evident in numerous tales of the hugely unexpected actually happening, from disease epidemics in crop monocultures changed at just a single gene, to global warming manifestations of the so-called butterfly effect.

I think we can only begin to come to terms with these potentially enormous implications of our actions by shifting intellectually, emotionally and practically, as deep ecologists urge, from egocentricity to ecocentricity, putting rationality in its place and giving precedence to our *living space*. To do this, we will need to develop different metaphors, language and philosophical and practical approaches from those that currently dominate rationalist thinking.

The immersive philosophy of 'inclusionality', which I am currently working on with others, is a response to these needs (Rayner, 2000a). This philosophy effectively views all things as dynamic contextual inclusions that both include and are included in space. The dualistic or discretist separation between insides and outsides, explicit contents and implicit context is subsumed within a reciprocally transforming dialogue, mediated across dynamic boundary interfaces transcending all scales of organization.

Water: the dynamic contextual medium of life, and its inclusional significance

For me, a wonderful metaphor for understanding the reciprocal relationship between explicit contents and actions, and implicit contextual containing space, can be found in river systems that both shape and are shaped by the landscape they flow through.

By taking substance out from their catchment, rivers effectively *make their own inductive space* – they both create and follow paths of least resistance. The same is true of all organic life forms and perhaps even all universal features that emerge and dissipate through the reciprocal dynamic relationship between inner and outer inductive holes – spelled H.O.L.E.S. If holism was about holes, incompleteness and the inductive influence of emptiness, rather than about wholes as entireties made up of more than the sum of their component parts, then I'd be all for it. It's *incompleteness*, not *completeness*, which *sustains* universal dynamics and our empathy for one another and our living space. There is nothing more repellent, more isolating, than self-sufficiency.

In organic life forms, as we know them here on earth, the *medium* through which this *reciprocal inductive relationship* between inner and outer holes, inner and outer space occurs, is none other than *water*. And indeed I think much of the *language of inclusionality* can be framed in terms of *the language of water*.

The way that water is channelled within dynamic living system boundaries to yield diverse patterns of environmental relationship is evident in the river-like pathways formed by these systems whenever they are observed *in context* rather than in individual snapshots of space and time. This is easiest with life forms that *grow* rather than *move bodily* from place to place and so have *indeterminate body boundaries* that *map* their own life history. However, it is also possible, given the appropriate imaging or imaginative processes, with those many animals that appear superficially to spend their lives as discrete individual units (see Rayner, 2000b).

I try to represent this relationship between life forms and flowing water forms in many of my paintings (for example, Figures 1.3–1.5).

Water is, and always has been, the receptive medium into and through which life forms gather and distribute the sources of energy that puts them in motion via processes of photosynthesis, chemosynthesis, digestion, respiration, transport and translocation. It provides the continuity between generations through and in which genetic information can flow and be exchanged and expressed in endlessly diverse forms (Rayner and Way, 1999).

The way in which water is arrayed and re-arrayed in living systems depends both on its own physical properties and those of the dynamic living system boundaries that retain it, and which change in versatile response to differing phases and circumstances of life, notably *deformability*, *permeability* and *continuity*. In the dynamic context of these changeable, water-retaining boundaries, life histories are not homogeneous storylines with discrete beginnings and endings, or even forever-closed life cycle circuits, but rather spiral trajectories wound around axes of time. Each turn of the spirals opens out from and then returns to a condensed, coherent, totipotent state in which all the creative possibilities of past, present and future are contained.

But what mechanism actually underlies this changing relationship at dynamic boundaries between insides and outsides? Here, I suggest that the involvement of water with another of the classical elements, fire, becomes important, particularly with respect to the availability of reducing fuel and oxidizing power, especially oxygen itself (Rayner, 1997, 1998; Rayner and Way, 1999; Rayner et al., 1999).

Oxygen might well be regarded as the world's first dangerously addictive drug, a Shiva-like activator and destroyer of life, which, when combined with reducing power delivered through the respiratory chain provides a boost of chemical energy in the form of ATP, but in excess results in the degeneration of protoplasm. It owes this duality of threat and promise to its affinity for electrons, which it receives one at a time in the course of its reduction to water, so generating highly reactive oxygen species and free radicals that can destroy the chemical integrity of cells. There are four main ways in which life forms respond to this duality. They may be 'gas-guzzlers', using up reducing fuel in dissipative self-differentiating structures until both they and their supplies become exhausted. They may protect their protoplasm internally by producing oxygen and free radical-scavenging 'secondary metabolites' and enzymes.

Figure 1.3 'Ivy River' (oil painting on board by Alan Rayner, 1997). An ivy river sweeps down from its collecting tributaries in steep-sided, lobed valley systems in high mountains, through dark forest and out across a sunlit, starkly agricultured, flood plain. Thence it delivers its watery harvest through deltas of leaves and fruits to a sea filled with the reflection of sunset. The fruits and leaves of a real ivy plant fringe the view of the distant river. The erratic pattern of veins in the lobed leaves contrasts with the focused pattern in the unlobed leaves and reflects the difference between the energy-gathering and energy-distributing parts of the river.

Figure 1.4 'Loving Error' (oil painting on board by Alan Rayner, 1998). This painting illustrates the dynamic interplay between differentiation and integration, irregularity ('error') and regularity, and negative draining and positive outpouring that is embedded in living system boundaries. The erratic fire in the venation of a lobed ivy leaf is bathed in the integrating embrace of a heart-shaped leaf which converts negative blue and mauve into positive scarlet and crimson. The midrib of the heart-shaped leaf emerges as a bindweed which communicates between extremes of coldness and dryness.

They may protect their protoplasm externally by forming sclerotized, oxygen-impermeable boundaries, actually incorporating oxygen into chemically cross-linked polyphenolic, lipid and proteinaceous compounds through the agency of peroxidase and phenoloxidase enzymes. They may actively or passively enable or allow degeneration to occur. These responses are of special significance in terrestrial habitats. Here, oxygen is capable of diffusing 10 000 times faster through the gaseous phase than through water, a fact that may well have been crucial in such fundamental processes as the heteromorphic alternation of haploid and diploid generations and the evolution of arborescent form (Rayner, 1997).

Figure 1.5 'Fountains of the Forest' (oil painting on board by Alan Rayner, 1998). This was painted for the British Mycological Society to depict the intra-connectedness of trees and fungi. Within and upon the branching, enfolding, water-containing surfaces of forest trees – and reaching out from there into air and soil – are branching, enfolding, water-containing surfaces of finer scale, the mycelial networks of fungi. These networks provide a communications interface for energy transfer from neighbour to neighbour, from living to dead and from dead to living. They maintain the forest in a state of flux as they gather, conserve, explore for and recycle supplies of chemical fuel originating from photosynthesis. So, the fountains of the forest trees are connected and tapped into by the fountains of fungal networks in a moving circulation: an evolutionary spiral of differentiation and integration from past through to unpredictable future; a water delivery from the fire of the sun, through the fire of respiration, and back again to sky, contained within the contextual boundaries of a wood-wide web.

Source: From Rayner, 1998.

Here, then, can be seen the fundamental way in which the contextual responses of life forms to the threat and promise of oxygen has shaped their evolutionary course. These responses cause boundaries to open, seal, fuse and degenerate, so enabling energy sources to be assimilated, distributed, conserved and redistributed in accord with local circumstances.

It is in this dynamic, watery, airy and fiery context, with which they co-evolve in reciprocal inductive relationship, that I think the true importance and meaning of genes can be fully appreciated. The occurrence and expression of genes both influences and is influenced by the dynamic properties of this context, much as the ingredients of a river both influence and are influenced by the river's course through its containing landscape. Genes are not and cannot be controllers and predictors of their own environmental destiny, and neither can we. The best that can be done by them and us is to *feel* our way, by attuning our inner space with our outer space through reciprocal relationships mediated at the dynamic boundary between the two.

Conclusions

So, what does all this mean for rationality and the way we should set about understanding how the world works and our own relationship to it? Clearly, the reality of environmental relationships is far more involved

Figure 1.6 'Oashiss' (oil painting on board by Alan Rayner, 1998). This painting depicts the vitality and unpredictability of the partnership between DNA and water, the informational traffic and the contextual waterways, of living systems. A riverine snake, with DNA markings, guards a waterhole in a desert of sand particles blown into waves. Pebbles at the edge of the water, modelled on the 'stone cells' ('sclereids') that make pears gritty, are separate, yet intra-connected via their cores. A goat skull and a fish out of water show the effect of exposure to dryness. How do we react to the snake? Do we attempt to control and predict its movements? Do we recoil from it? Do we relate ourselves to it? Which of these reactions promises most, or most threatens our quality of life?

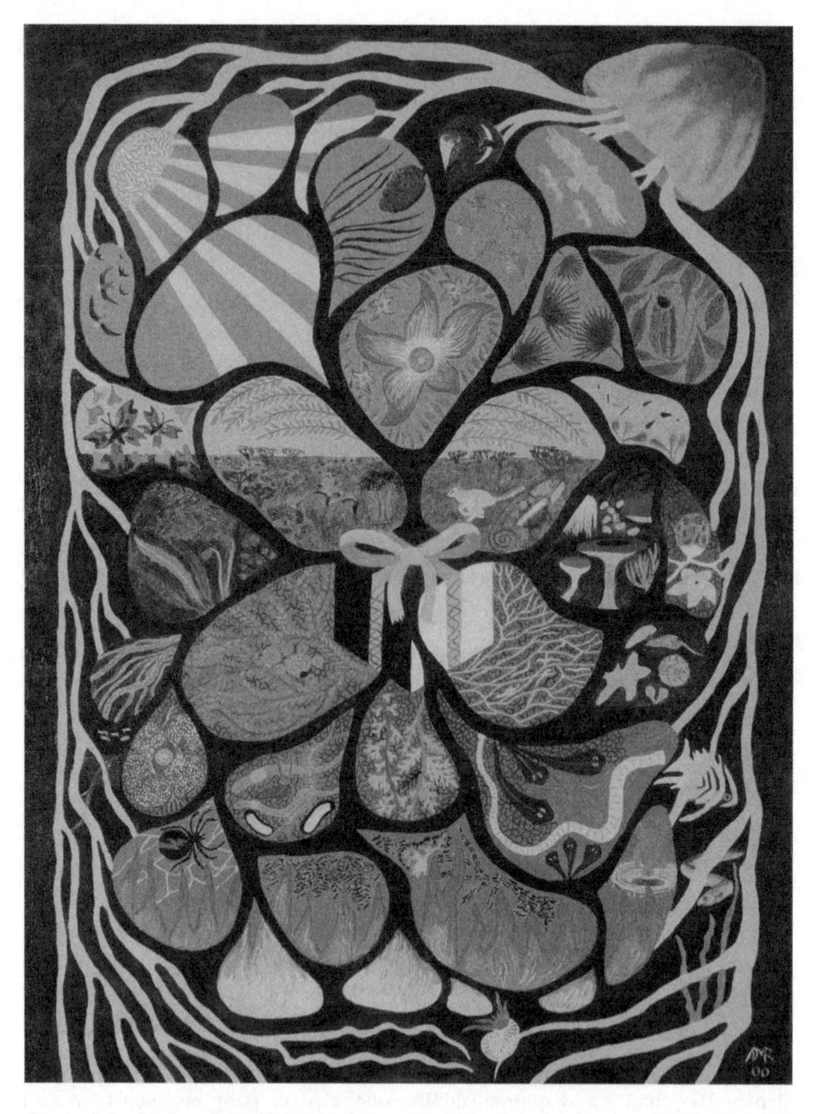

Figure 1.7 'Future Present' (oil painting on canvas by Alan Rayner, 2000). The gift of life lies in the creative infancy of the present, whence its message from past to future is relayed through watery channels that spill out and recombine outside the box, reiterating and amplifying patterns over scales from microscopic to universal.

than rational methods of enquiry can hope to account for in anything more than the very short term, and there is a need to obtain a much more all-round view. This does not necessarily mean, however, that we should discard rational enquiry altogether. Indeed, once we have acknowledged its limitations as not being 'all there is', I think we can *value* its explicit, laser-like focus as a very useful high resolution tool. This tool, when *complemented*, through *dialogue*, by the collective imagination and insights arising inclusionally from many, diverse perspectives, can then help illuminate specific aspects of a much larger implicit, holographic reality. In this way we can stay attuned to the implications of our dynamic living space, rather than continue to create dissonance by assuming control over what our restrictive analytical vision prevents us from seeing.

These conclusions are reflected in two more paintings (Figures 1.6 and 1.7).

Towards an environmental research agenda: combining artistic and scientific perspectives of implicit space and explicit information

Perhaps, most fundamentally, what I am suggesting is that in order to attune more empathically with one another, other life forms and our surroundings, there is a need for a radical reorientation in the way we think about the nature of *contextual space*. To achieve this will involve more than tinkering with our methodologies and ways of seeing. It will involve a profound, and for many people disturbing, upheaval of the intellectual landscape our rationalistic vision has conditioned us to believe in. In the words of novelist Lindsay Clarke (pers. com.), it will involve nothing less than a 'transformation of consciousness in our time', a 'shift away from the fissive mythology of positivism back towards a lively sense of the sacred', a sense of *Place*.

This is where I think our future environmental research agenda can benefit especially from a reintegration of scientific and artistic perspectives. For Science and Art, as they have both lost touch with one another and reality, have *abstracted* Knowledge *content* out of *spatial* Context in divergent, but ultimately complementary ways. On the one hand, the fixed reference frame of rationalistic Science has tended to treat space as an insubstantial and consequently passive *absence*, whose dynamically transforming shape can safely be *excluded* from any consideration of the assertive material properties of *explicit things*. On the other hand, Art has increasingly excluded the explicit aspect of things in order to explore an implicit space in which *anything goes*. Whereas

Science has left no room for imagination by getting too much of a grip on explicitly packaged bits and pieces, and so mistaken the reference frame for the whole picture, Art has loosened the frame so much that it can't keep hold of the picture.

The tremendous prospect of bringing art into science and science into art is that by so doing we can appreciate the reciprocal interdependence and consequent inseparability of implicit *contextual space* and the explicit *information* that gives heterogeneous expression to that space *in the form* of features. Here, insides are not *sealed* forever within the boundaries of outsides. Things are not physically discrete *bodies*, isolated by space, nor even are their outsides all *inter*connected by some explicit external *web* of material presence. Rather, they are *embodiments* of that implicit space which is not the physical absence that separates them, but rather the *labyrinth* of immaterial, non-resistive, inductive (that is, super-conductive) presence that *intra*-connects them by uniting their insides *through* gaps in their boundaries to their outsides.

To try to get some more feel for what this means, try to imagine a world or universe with no space. Is there any *possibility* for movement or distinctiveness? Now try to imagine a world or universe of pure space. Is there *anything* there? For me, the conclusions from such imaginings are inescapable. Space is pure, implicit, insubstantial *possibility*, but for that possibility to be *realized* – expressed in distinctive, heterogeneous features – it has to be given shape, that is *informed*, by something explicit. Gregory Bateson alluded to this explicit something as 'the difference that makes a difference', *information*. However, by the same token, this *information* without contextual space is meaningless, makes no difference and has no possibility for independent expression.

Explicit information and implicit space are therefore both inseparable and dynamically co-creative. They make and are shaped by the other in the same way that the water in a river system makes, shapes and is shaped by the space through which it flows, as it erodes rock and deposits sediment. And the making of space makes possible a flow that makes more space – an 'autocatalytic flow' – as when people walking across a meadow create and consolidate an inductive path by following their leader.

This inclusional view of information as content in relation to spatial context contrasts with the discretely packaged informational units of rationalistic, binary (either/or) logic and digital computers. Inclusional information, far from being broken up into transmissible bits and pieces of pure machine *code* that need to be protected from contamination by 'outside interference' or 'noise', produces vibrant, flexible *language*. It *folds* into and around the space it relates to as a dynamic matter–energy-containing boundary that nests inner spaces within outer spaces across all scales from subatomic to universal. This boundary is not the fixed limit of particulate things – it does not *define* – but rather provides the *mediating surface* or *interface* through which inner and outer spaces reciprocally and simultaneously transform one another.

So, the Big Story of Life and the Universe is the 'Hole Story', not the 'Whole Story'. To be dynamic, things consist of 'informational holes' – *lined spaces* – not wholes and parts complete, and so static, within themselves. These holes are inductive, attractive – they have pulling power: the beauty of a cathedral is in the space that its walls line, not in those walls alone. And the holes puncture the rationalist's box that has held us like Schrödinger's Cat in secure paradoxical bondage, longing to escape into the real world where inner space connects with outer. And, if there is anything on earth that can find these holes and show them to us for what they are, we need not look for anything rare. We need only to regard that overlooked, taken for granted *commonplace* – water, the dynamic contextual medium without which the genetic code of DNA could not be translated into the informational surface that co-creates the diversity of life itself.

To pursue a more inclusional approach to environmental relationships, which combines artistic *imagination* with scientifically derived *information*, represents a tremendous challenge. It will test to the utmost our willingness to encompass *diverse* viewpoints so that they *complement* rather than *conflict* with one another, and so facilitate truly co-operative inquiry, of the kind discussed by Kent (2000). We will also have to become wise to the shortcomings of some of our most cherished and trusted rationalistic concepts and methods of inquiry and problem solving and, where necessary, adopt new approaches. Progressing from rational to inclusional logic demands that we rethink our use of such fundamental calculational tools as binary numbers, probability theory, infinity and infinitesimality, and their application to understanding space, time, energy, matter and their evolution. It demands that we finally let go of the notion of independence and come to regard uncertainty and creativity in terms of *possibility* rather than *probability*, and evolution in terms of relational transformation and natural inclusion rather than survival of the fittest and natural selection. We will also need to develop systems of numbers that can actually *relate* to one another (as with the ternary fluid logic numbers of Shakunle, 1994). All that's going to require a lot of co-operating and a lot of listening. If we don't manage it, the cumulative damage to our living space may become irrevocable. If we do, we may truly fulfil our potential to deserve the name of Human*kind*.

Acknowledgements

Many people have participated in the exchange of ideas that has led to my writing this article. I would like, however, especially to mention members of two discussion groups. These are the 'Inclusionality' group, notably Ted Lumley, Doug Caldwell, Dirk Schmid, Lere Shakunle, Martine Dodds, Sidney Mirsky, Seb Henagulph and Songling Lin, and Bath Bio*Art group, notably Caroline Way, Sandi Bellaart, Geoff Abbott, Jeff Schmitt, Linda Long, Kevan Manwaring and Christian Taylor. I would also like to thank one of my project students, Juliet Muir, whose artwork, 'Human Kind?', so beautifully portrays the paradox of rationalism and the holes through which we may pass beyond it to a richer relationship with nature.

References

Kent, J. (2000) 'Group inquiry'. In C. Truman, D.M. Mertens & B. Humphries (eds), *Research and Inequality,* pp. 80–94. London: UCL Press.

Rayner, A.D.M. (1997) *Degrees of Freedom – Living in Dynamic Boundaries*. London: Imperial College Press.

Rayner, A.D.M. (1998) 'Presidential address: fountains of the forest – the interconnectedness between trees and fungi'. *Mycol. Res.*, 102, 1441–9.

Rayner, A.D.M. (2000a) 'Put genes in context – life is not like a box of Lego'. *Times Higher Educational Supplement*, 28 November 2000.

Rayner, A.D.M. (2000b) 'Challenging environmental uncertainty: dynamic boundaries beyond the selfish gene'. In A. Warhurst (ed.), *Towards an Environment Research Agenda,* vol. 1. London: Macmillan.

Rayner, A.D.M., Watkins, Z.R. & Beeching, J.R. (1999) 'Self-integration – an emerging concept from the fungal mycelium'. In N.A.R Gow & G.M. Gadd (eds), *The Fungal Colony.* Cambridge University Press.

Rayner, A.D.M. & Way, C. (1999) 'Evolutionary waterways: the contextual dynamics of biological diversity'. *Frontier Perspectives*, 8(2), 33–7.

Shakunle, L.O. (1994) *Spiral Geometry. The Principles (with Discourse)*. Berlin: Hitit Verlag.

Further reading

Bate, J. (2001) *The Song of the Earth*. London: Picador.

Blatter, J. & Ingram, H. (2001) *Reflections on Water – New Approaches to Transboundary Conflicts and Co-operation*. MIT Press.

Boyle, D. (2000) *The Tyranny of Numbers – Why Counting Can't Make Us Happy*. London: Harper Collins.

Laszlo, E. (1996) *The Whispering Pond – A Personal Guide to the Emerging Vision of Science*. Shaftesbury, Dorset: Element Books Ltd.

Rothenberg, D. & Ulvaeus, M. (2001) *Writing on Water – A Terra Nova Book*. MIT Press.

2
Inclusionality–Exclusionality: Environmental Philosophy and Simulative Politics

Ingolfur Blühdorn

Summary

In direct response to Alan Rayner's *Inclusionality – An Immersive Philosophy of Environmental Relationships* in this volume, this chapter critically reviews the validity of some stereotypically reproduced arguments within the environmental debate. It aims to demonstrate how late modern societies have moved beyond the philosophy of inclusionality that provided the foundation for ecologist thinking and politics. The *post-ecologist* realities of contemporary society however, are, arguably, obscured by a societal practice that is described as *simulative politics*: whilst there is no serious confidence – nor actually ambition – that the modernist project and promise of ecological thought will ever be completed, late modern society keeps reproducing the illusion that the ecologist ideals are still valid and on the agenda. Academics and intellectuals can hardly avoid contributing to this collective strategy of simulation, or they expose themselves to charges of having abandoned humanist values, and fatalistically taking refuge in apologias of an unacceptable status quo. Trying to pierce this protective screen surrounding late modern consciousness, this chapter aims to expose how the well-intended appeal for an *immersive philosophy of inclusionality* has become a function of an individualized reality of exclusionality.

Introduction

Having to abandon long established and cherished beliefs tends to be painful, and there is risk and uncertainty in every such departure.

Nevertheless, if we allow ourselves to get unduly attached to established perspectives, our perceptions of contemporary society and our recommendations for its further development can easily turn stale and rather unbearable. This is the case for certain currents within the ecological debate. Whilst economic development and growth have once again clearly taken priority over any concerns about climate, resources or biodiversity; whilst eco-political negotiations and agreements proceed at the pace and to the conditions of the slowest and most refractory countries (which are often at the same time the most affluent and ecologically unsustainable ones); whilst the idea that an ecological rationality genuinely differs from economic thinking has been abandoned and ecological issues are being rephrased as managerial issues and issues of resource efficiency; whilst Green Parties are trying to secure their survival by redefining themselves as agents of civil society and consumer safety; whilst eco-movements are fragmenting into short term, often radical and violent, single issue campaigns lacking any positive vision, continuity and mass support; whilst environmental NGOs are mutating into global businesses and eco-service providers; in other words, while we are confronted with huge transformations in the way ecological issues are being formulated and dealt with, significant parts of the ecologically committed literature – Alan Rayner's chapter in this volume being a good case in point – keep reproducing categories and appeals which have long since lost any analytical validity and political potential.

Environmental philosophers and ecological idealists are still talking of our *attitude problem*, of the urgent need for a *radical reorientation* in the way we think. We are urged to appreciate the implications of our exclusive, exploitative and domination-orientated scientific rationality. We are told about the *real happiness and fulfilment* that we forsake and the cumulative damage to our living space that may become irrevocable if we do not manage to reorganize our lives and societies on the basis of a new *immersive philosophy of inclusionality*. This lamentation, these promises, these appeals have been regurgitated ever since the very beginnings of ecological concern. They echo concepts, categories and ideas which have their origins in the European movement of the Enlightenment and have been shaped and reshaped by thinkers like Rousseau, Kant, Marx, Weber and a plethora of twentieth-century proponents of critical socio-political theory. However, in *late* modern or even *post*-modern affluent societies, neither these appeals nor the underlying concepts and analyses have ever been convincing. Unsurprisingly, their effect has always remained rather limited. After more than a century of eco-political analysis and campaigning – and in particular against the background of the radical transformation which late modern societies have experienced since the end of the Cold War and the beginning of the era of globalization – it is therefore about time to acknowledge that these well-rehearsed arguments have exceeded their 'best before' date. They are theoretically underdeveloped, sociologically blind and normatively dubious. And from an ecological perspective this old-style lamentation and rhetoric may actually even be regarded as dangerous and counter-productive: what was once critical and revolutionary, has today features of a social tranquillizer stabilizing the very conditions ecologists were trying to change.

This chapter argues that rather than continuing the futile appeals for a radical change in our thinking, attitudes and patterns of societal organization, eco-political thought itself has to undergo radical change. Rather than repeating the stale pleas for a *participatory philosophy of inclusionality* or an *immersive philosophy of environmental relationships*, ecological thought has to take account of the radical social and cultural change that has actually occurred in late modern societies and that has to be acknowledged as the starting point for any eco-political analysis and strategy. In as much as environmental sociology – contrary to environmental philosophy and ecologist ideology – is sociologically descriptive rather than ecologically prescriptive; in as much as it retains a critical distance from established ecological imperatives and expectations of ecological correctness, environmental sociology is uniquely positioned to take a leading role in this reorientation. This chapter aims to contribute to this critical review of the ecological debate firstly by exploring the contemporary meaning and significance of the principle of *inclusionality* and secondly by developing an analytical framework for the interpretation of contemporary discourses of inclusionality as *simulative politics*.

The question that is to be addressed here is why the old ecologist demand for a new socially and ecologically benign philosophy of inclusionality is so difficult to implement in empirical reality. The answer that is being suggested is that contemporary society is essentially an exclusive society. Correctly Jock Young noted that 'the transition from modernity to late modernity can be seen as a movement from an inclusive to an exclusive society. That is from a society whose accent was on assimilation and incorporation to one that separates and excludes' (Young, 1999, p. 7). And arguably, the problem is not only that the principle of exclusionality is so deeply rooted in contemporary society that any appeals for a reorientation of the way we think and behave necessarily must remain pointless, but more importantly, late modern society does not really have any genuine ambition to become an inclusive – and ecological – society. This holds true even though in recent years there has been a lot of public debate about social inclusion, civil society and environmental sustainability, and even though there are plenty of policy programmes aiming to put these goals into practice. In order to explain and justify these hypotheses, two main issues need to be discussed: (a) In what respect and to what extent may contemporary late modern society legitimately be described as an exclusive society that has abandoned the principle of inclusion and is not interested in its reintroduction? (b) How does this hypothesis that the modernist principle of inclusionality has been – or is being – replaced by the genuinely post-modernist principle of exclusionality fit together with the official rhetoric and public policies of social inclusion.

At least at a preliminary level, the latter of these two questions is the easier one to answer. The apparent contradiction between the assumed structural principle of exclusionality and the ubiquitous rhetoric of inclusionality can be explained by the thesis that in the context of late modern society the concept of inclusionality has changed its meaning – almost into the opposite of what it once implied. Beck, Giddens and a range of others have suggested that societal development reflexively undermines and destroys the foundations of modern society and that this process of *reflexive modernization* (for example, Beck, Giddens and Lash, 1994; Beck, 1997; Blühdorn, 2000b) is radically reshaping contemporary societies. Reflexive modernization means that the established certainties and beliefs of traditional modernity are being deconstructed and reconstructed in a different way. This affects both the established structures of modern society as well as the sociological concepts and categories for its (self-)description. The concept of inclusionality, too, can be said to be undergoing such a process of reflexive redefinition. Whilst the concept itself is being retained, its meaning is changing in such a way that the politics of inclusionality essentially amounts to a politics of exclusion. In order to fully explain this hypothesis, and as a preliminary analysis providing the foundations for the discussion of the practices that were described as simulative politics, this deconstruction and reconstruction of the concept of inclusionality shall now be explored in some detail. The guiding questions for this analysis are: Inclusion *of what*? Inclusion *into what*? Inclusion *for what reason or purpose*?

The modernist understanding

The modernist idea of inclusionality has its origins in the European movement of the Enlightenment and, in particular, in idealist Enlightenment thought. The most fundamental value of modernity, the value that claims universal validity and that is attached to any human being is undoubtedly *freedom*. Freedom emerges from the fact that human beings command rational faculties which allow them to think and act in ways which are not predetermined by nature, that is, by uncontrollable instincts and desires. This capability to decide and act independently from, even in contradiction to, the forces of nature distinguishes the human Self from the Other and provides the basis for the formation of human *identity*. From a modernist perspective, freedom and rationality do not only tie human individuals into a community integrating the entire human species, but in as much as Enlightenment thinkers believed that nature was organized by, and functions in accordance with, the laws and rules of that same rationality, human beings were part of an *all-embracing*, a universal rational context.

Inclusionality in this sense is inclusionality without exclusionality. Very importantly, inclusionality in this sense means first and foremost that integration and participation, rather than having to be achieved, have always been the case. Nevertheless, because inclusionality, because the degree of participation in rationality and freedom, can be developed and increased, particularly the idealist Enlightenment tradition has always understood inclusionality as a project, the unfinished *project of modernity* (Habermas, 1984, 1987a,b). Whilst human individuals could cultivate their rational faculties, transcend restricted individual rationality that was distorted by self-interest and thus get ever closer towards genuine freedom and absolute reason, the sciences could progressively uncover how the perceived secrets and mysteries of nature had fully rational explanations. Any progress in human rationalization and scientific discovery would thus advance the project of realizing (recognizing and developing) inclusionality, the all-embracing coherence of reason. In response to the question *Inclusion into what?*, we may thus note that the modernist Enlightenment notion of inclusionality means the (increasing) awareness that we are part of – that the entire universe represents – an all-embracing rational coherence.

In response to the question *Inclusion of what?*, the Enlightenment notion of inclusionality can be said to imply inclusion of the human being *qua* participant and representative of absolute, transcendental reason. As rational beings, human beings are free and equal. They are autonomous, however they are not autonomous as individuals, but primarily as a species. As individuals, human beings are free and autonomous only to the extent that their restricted individual rationality participates in absolute reason. However, empirically individual rationality is distorted by all kinds of natural instincts and desires and thus fairly remote from absolute reason. This implies rather narrow limits for any individual's claim to inclusion into the realm of freedom and rationality.

The subject of genuine autonomy is, if anything, the human species as a whole, or to use Kant's term, the *transcendental subject*.

The Enlightenment belief in, the modernist project of, inclusionality has always implied an imperative, which was at the same time a moral, that is, categorical, imperative and an imperative of human prudence. The moral imperative emerges from the idea that, as Kant put it in his stunningly contemporary *Idea of a Universal History with a Cosmopolitan Purpose* (1784), 'nature has willed' that 'all the natural capacities of a creature are destined sooner or later to be developed completely' (First Proposition). The human capacity of freedom and rational autonomy, in particular, can, as Kant states, 'be fully developed only in the species, but not in the individual' (Second Proposition). Inclusionality is therefore imperative, and humankind has to keep working towards resolving all tensions in the empirical world and reconciling all social antagonisms in order to bring about the, eventually cosmopolitan, 'perfect political constitution as the only possible state within which all natural capacities of mankind can be developed completely' (Eighth Proposition).

Apart from being imperative for 'the realization of nature's hidden plan' (Eighth Proposition) for an all-inclusive rational system which will, at some distant stage, provide the context for each constitutive element to develop its fullest potential, the principle of inclusionality was also assumed to have immediate short-term benefits. Subjecting oneself to the rule of reason, that is, renouncing the original freedom of the Hobbesian *state of nature* and entering a *social contract*, provides protection against incalculable behaviour by fellow humans and creates a space for self-determination and self-realization. Subjecting nature to the rule of reason, that is, discovering by means of the natural sciences its rational laws and principles, makes nature calculable and controllable and takes away its mysteriousness and threat. The function or purpose of inclusionality, of integrating oneself and every phenomenon of nature into the all-embracing systematic rational coherence is thus to increase *negative freedom*, that is, freedom from external threats and uncertainties, and to maximize *positive freedom*, that is, self-determination and self-realization. The principle of inclusionality provides the framework within which human autonomy and self-realization, that is, the formation of human identity, thrives best. Very importantly however, this modernist notion of inclusionality regards autonomy, identity and self-realization primarily as *social* rather than *individual* categories.

The analytical distinction between inclusionality as a *categorical imperative* and inclusionality as an *imperative of prudence* is important because it reveals how the principle of inclusionality may actually become counter-productive. If inclusionality is not an intrinsic value but instrumental for the realization of a conceptually different end, other means might turn out to be more appropriate. In idealist Enlightenment thought such a conceptual distinction between means and ends did not make much sense. However, if the goals of autonomy, identity and self-realization were to be defined differently, it might become an imperative of prudence to abandon the principle of inclusionality. For the further discussion of the concept of inclusionality it is therefore essential to explore how the key parameters in the above analysis – freedom, equality, rationality, participation, identity, autonomy, self-realization – have changed their meaning. This will reveal how the idea of inclusionality itself has changed. The questions *Inclusion of what? Inclusion into what? Inclusion for what purpose and on what normative grounds?* will remain at the centre of attention.

Inclusion and colonization

A major problem for the modernist project of inclusionality originated from the fact that rationality did actually not have the kind of metaphysical anchorage Kant described as transcendental reason. On the one hand modern society continued – and continues to the present – to maintain the existence of universal values and the possibility of the all-integrating, all-embracing, rational system – global civil society. On the other hand, however, implicitly acknowledging its transcendental deficit, that is, the lack of sound normative foundations, modern society, from a very early stage, institutionalized as its core constitutive principle the plurality of separate rationalities which were meant to establish a system of mutual checks and balances. Modern society is based not only on the separation of powers (legislature, judicature, executive), but in addition to this, on the independence and integrity of realms like religion, economy, law, science, politics, education and the media. A pact of non-interference is supposed to regulate the autonomy of each of these realms, which all fulfil indispensable functions for modern society, and which all depend on the integrity of their specific logic or *code* (Luhmann, 1995). Despite its belief in the principle of inclusionality and the singularity of reason, modern society is thus based on the principle of irreducible plurality which is supposed to be integrated by the relationship of each independent realm to the collective subject, that is, to society (not the individual!), for which it fulfils a specific function.

This plurality of different logics however, that is, the fact that the economically profitable may contradict the legally acceptable, that the legally correct in turn may differ from the morally acceptable, that the scientifically correct may conflict with the politically correct or the politically feasible and so on, gives rise to a competition of power between the different rationalities or function codes. There is the threat of one rationality infringing upon the integrity of another, of one *colonizing* others – as is most evidently the case with economic rationality which in contemporary society has permeated and transformed all other realms, making it ever more difficult to maintain the belief in their integrity and independence.

In addition to this, the function systems tended to develop their own necessities and dynamics, putting not only their mutual integrity under threat but also raising doubts about the extent to which they really *served* society rather than beginning to dominate it. At least partially, the function systems soon emancipated themselves from their instrumental role and adopted the status of a purpose in themselves. Particularly from the perspective of the individual, the function systems turned into a threat, a source of domination. Habermas' well known distinction between the function-centred *system* and the individual-centred *life world*, with the latter being colonized by the former (Habermas, 1987a) sought to capture exactly this development. Against the background of these three threats, that is, the systems invading each other, the systems beginning to dominate rather than serve society, and systemic necessities jeopardizing the integrity of the individual, inclusionality turned into a problem. Further developing Marx's critique of capitalism, twentieth-century social theory – most notably Horkheimer, Adorno, Marcuse, Habermas – thus began its critical discussion of the *wrong kind of inclusionality*, that is, inclusionality that reduced rather than increased freedom, that was forced and implied colonization, domination and alienation rather than promoting the project of modernity, that is, rational autonomy and the full development of human capacities.

What was required was a new and different effort of emancipation – this time not from the pre-rational sources of domination, that is, the forces of nature within the human being and in its physical environment, but from the dominance of self-serving social systems which had been established by human beings but had then grown out of control. Critical social theory identified a radical contradiction between the inclusionality Enlightenment thinkers had envisaged, namely the inclusionality of universal values and the all-embracing liberating rational context, and the *delusive context* (Adorno) of exploitative *instrumental rationality* that had factually installed itself instead, and that was threatening to become universal. Continuing the Marxist tradition, the critical agenda was therefore to prevent the universalization of the wrong kind of rationality and values, to erect barricades against the seemingly unstoppable spread of false consciousness. The memory and the ideal of the non-exploitative, non-instrumental, liberating rationality that was expected to provide the most suitable framework for the free development of the natural capacities of both human beings as well as non-human nature had to be preserved. Nature, the victim of the wrong kind of inclusionality, had to be protected. Ecologist thought, with its demand for a radical review of the dominant form of exploitative, destructive, instrumental rationality, was firmly embedded in this intellectual context. Its demand for the liberation of inner and outer nature, for the acknowledgement and preservation of nature's integrity, for a participatory philosophy of inclusionality and an immersive philosophy of environmental relationships, could thrive in this climate and can only be understood within this context.

Reflexive redefinition

In the second phase of emancipation – in order to avoid the wrong impression of two consecutive phases it might be more suitable to speak of two overlapping layers of emancipation – the *individual* acquired special significance and adopted a special responsibility. In as much as *society*, societal function systems, represented the irrational, the source of domination, exploitation and destruction, the individual was required to distinguish itself from society, to position itself *vis-à-vis* society, and protect or cure society from its deformations. The individual thus became the *locus*, the source, the subject, of values providing the normative measure for the critique of society. Whilst first-order emancipation, that is, the emancipation of rational beings from the pre-rational forces of nature, had been a project for humanity at large, second-order emancipation thus became a project for the individual which aims to emancipate itself from false social consciousness and social practices, and, in particular, from the restrictions and domination emanating from society's function systems. The individual was now required to establish its own individual identity which was expected to function as the source of the corrective norms curing the malformations of society. The individual became the ultimate point of reference. Freedom, equality, rationality, participation, identity, autonomy, self-realization, and so on, which in idealist enlightenment thought had all been defined collectively, that is, with reference to humanity as a whole, were now being redefined with reference to the individual. And this redefinition of its constitutive parameters, obviously, did not leave the meaning of inclusionality unaffected.

The most crucial implication of this enthronement of the individual is the reversal of the relationship between the universal and the particular: whilst in the idealist tradition the particular individual had been understood to participate in (to be included in) transcendental qualities and universal values attributed to humanity *qua* humanity (but certainly not directly to the individual), the emphasis now shifted towards qualities and values attached to concrete individuals which were supposed to be universalized. Universalized individuality takes the place of individual participation in universality. And to the extent that the individual now positioned itself *vis-à-vis* society, that it sought to distinguish itself from society in order to generate the normative yardstick for the critique of the latter, the principle of inclusionality – as the main pattern of identity formation – was replaced by that of exclusionality. For this reason, the enthronement of the individual did not only imply that individual inclusion into universal values gives way to universalized individuality, but to the universalized principle of exclusion. The meaning and implications of this shift, the way in which the constitutive parameters of the idea of inclusionality and thus inclusionality itself have been redefined, need to be explored in more detail. The particular problem of such an analysis is that the various concepts and their meaning are tightly woven into each other requiring the investigation to proceed in circular movements.

In contrast to humanity's emancipation from nature, which was based on universal values (primarily freedom), the emancipation of the individual was based on individual rights (primarily the right to self-determination). Incrementally, these rights were – and still are – being extended to ever larger social groups, until eventually they are expected to become universal. Whilst the individual emancipates itself from false, forced, restrictive, irrational inclusion (religion, tradition, self-serving function systems), it strives for a different kind of inclusionality, namely the generalization of the right for self-determination and self-realization. In the process of second-order emancipation, this newly defined inclusionality becomes the goal of societal progress and development, and the normative measure for the critique of modern society. The Self, the identity, that is to be established and realized is now, firstly, an individual rather than a collective self (identity); and secondly, it is constituted as the subject of empirical rights and demands rather than *a priori* qualities and values. This second aspect, in particular, implies a shift from the subject having its substance in itself, from the subject that might be described as self-sufficient, towards the subject depending – for the purposes of its self-realization, self-expression and self-experience – on the empirical world.

However, perhaps exactly because of this *externalization of the subject*, second-order emancipation could neither stop the economic system colonizing society's other function systems, nor could it prevent the colonization of the Self and its life world by the system. Individuality, individual identity, subjectivity as the possession of individual rights could most easily and effectively be experienced, expressed and acknowledged in form of the possession of material goods. The universal right to self-development and self-expression materialized most visibly as the universal right to (demand for) material consumption. Rapid economic growth during the first post-war decades provided the basis for this pattern of identity formation to take root. Autonomous self-development, self-realization, adopted the primary meaning of ever increasing product choice, of free consumer decision. The freely developing and socially valued Self took the form of the consumer ego. Individuality turned into the unique consumer profile. The contemporary individual thus constitutes and experiences itself in the act of consumption. In a societal framework in which the economic code has permeated and reshaped all human experience and social relationships, and which leaves no equally powerful criterion and measure of distinctiveness, exclusiveness and thus identity, material inequality and exclusion are becoming indispensable. Inclusionality in this context means participation in the system of material production and consumption. The demand for inclusionality thus turns into the demand for unrestricted participation (inclusion) in exactly that context, from which post-Marxist critical theorists – as well as ecologists – had sought to emancipate themselves: the capitalist growth economy and its dimensions, that is, training and qualification, the labour market, career opportunities, social benefits, and so on.

Crucially important in this context is that individuality, the redefined individual, which was meant to function as the source of true consciousness and of corrective values stopping the spread of the *delusive context* and curing modern society from its *pathologies*, no longer has any substance to set against the status quo. The earlier demand to resist *colonization* by the alien system, suddenly turns into the demand to be included into this system – in order to reproduce its inherently exclusive structure. This reversal has become possible and acceptable not just because the notion of identity has changed, that is, identity has contracted into the economic realm with genuinely non-economic dimensions of identity formation and expression having been radically devalued, but also because the economic system itself has changed in such a way that being integrated into it, absorbed by it, is no longer negatively experienced as colonization, as restriction or loss of freedom, but positively as the achievement of new freedom and opportunities.

The Fordist system was a system of standardization, massification and uniformity. It was undifferentiated and inflexible, leaving no space for plurality, distinctiveness and individuality. In contrast to this restrictive system, the contemporary global economy is highly differentiated and flexible, offering an almost infinite range of choices and thus options for identity formation and self-expression. Indeed the range of options is so overwhelming that the individual does not become aware that what the market offers are merely product options and consumer choices. One dimension of identity construction is thus fully developed at the expense, almost exclusion, of any other dimension. What emerges is a fairly one-dimensional identity, yet, an individual that does not claim (or have) other dimensions beyond the economic one, will not feel restricted. It can fully develop its freedom and its potentials within the existing (economic) framework. And in the unlikely event that any concerns of one-dimensionality, colonization or alienation should emerge – this will be discussed in more detail in the next section – contemporary society has developed powerful strategies to reassure itself of the existence and centrality of individuality, and to disguise the compression of the idealist autonomous subject into the consumer profile. The rhetoric of the unprecedented chances and opportunities emerging in the process of globalization, and the constant emphasis of the economy that its products and services are tailored to the particular needs of unique individuals are effective drugs against any such concerns.

In the all-embracing economic system *freedom* is thus being newly defined as *freedom of choice*. Freedom of choice differs from freedom in the enlightenment sense in that the latter implied the realization, the expression, the implementation of pre-existing values and qualities, that is, of a pre-existing identity. Freedom was thus an original, creative, generative potential. In contrast to this, freedom of choice is merely consumptive. It does not itself generate or shape structures and items, but all options are externally determined and presented to the subject of choice (consumer) which can merely accept or reject but has little or no influence on the range and quality of the options available. Freedom of choice, of course, soon reveals itself as pure contingency and arbitrariness. If the subject of free choice does not have clear criteria on the basis of which choices can be made, an extensive range of options turns into a source of disorientation, insecurity and fear. Against this background, externally defined necessities become desirable, or non-rational momentary impulses and inclinations fill the normative gap. Unless there is a firmly established identity as source of criteria for the management of multiple options, freedom of choice thus gives rise to the demand for external guidance, for leadership, for authoritarian rule, which all imply the opposite of individual freedom and autonomy. The attempt, however, to use an identity that has been established on the basis of consumer choices as the source of criteria managing further consumer choices is self-referential and circular; it turns choice into arbitrariness and identity into contingency. This sheds doubts on the very concept of identity, suggesting that in the context of late modern society the formation of identity in the traditional sense might no longer be possible. Also, it poses the question whether the process of globalization with its radically extended range of opportunities and choices implies a gain or a loss in terms of freedom. These questions take us deeper into the analysis of the reflexive redefinition of identity and freedom. Their investigation is prerequisite for understanding the post-ecologist constellation that is the focus of the Introduction.

Identity formation is a process that depends on repetition, ritual and stability. In the absence of *a priori* values and criteria, constant repetition establishes (essentially contingent) elements of, and experiences in, the empirical life world as the certainties and criteria on the basis of which individual identity may be constructed. These repetitions refer to the spatial or territorial dimension, to the objects of experience and to the normative rules regulating the relationship between the individual and its physical and social environment. They include the totality of culture-specific values, experiences and practices against the background of which individual characteristics can develop and identity emerge. The formation of identity thus demands the range of experienced variations to be narrowly restricted. In order for repetition to occur and thus identity emerge, territorial mobility, normative flexibility and the plurality of empirical experience need to be kept within tight limits, with a widening of the horizon occurring only at a pace that retains a balance between the certainty and security emerging from the fledgling identity and the uncertainty, risk and disorientation that emerges from difference. Arguably, an increasingly globalized, mobilized, pluralized, flexibilized and virtualized world does no longer provide the stability, steadiness and repetition that is required for the formation of identity. Fluid or *liquid* modernity (Bauman, 2000) is essentially incompatible with – outright hostile to – the idea and concept of identity. With its categorical demand for flexibility and innovation, with its idealization of constant change and life-long learning, late modern society has established structures and adopted value preferences representing the very opposite of identity. Of course identity is not a value in itself. Contrary to well-established sociological beliefs, the decline of the individual and identity does therefore not necessarily imply any experience of deficit or loss; nor does it mean that the concept of identity has been or needs to be completely abandoned. What this does suggest however, is that in late modern society the term identity must have radically changed its meaning.

As regards the question whether the process of globalization implies a gain or a loss in terms of freedom, it is evident that by participating in increasingly globalized function systems the individual expands its life world and range of experience. This seems to imply a radical increase in the range of opportunities for individual self-construction, self-development and self-expression, and thus a gain in terms of freedom. On the other hand, to the extent that in each of these increasingly comprehensive function systems the individual has to adapt to systemic imperatives and rules dictated by the logic of the respective system, the individual is not autonomous (free) and in control of its options. Furthermore, with the expansion of the function systems in which the individual participates, the individual's life world becomes increasingly complex and confusing. The more comprehensive and complex the life world becomes, the stronger and well established a subject and identity is required to integrate it. However, as was argued above, late modern society is not conducive to identity formation and breeds only weakly developed subjects whose integrative capacities are not strong enough to maintain the unity of an ever expanding life world. For this reason the individual-centred life world becomes amorphous and disintegrates. Yet if the process of globalization implies that both the individual as well as the individual-centred life world fragment and lose their distinctiveness; if there is no self to be realized, and if freedom of choice turns into arbitrariness, this appears as a loss rather than a gain of freedom and opportunities of self-development.

Despite all this, that is, despite the redefinition of freedom as freedom of choice and the dialectic conversion of radically increased opportunities into arbitrariness and indifference, there is not much space for any experience of restriction or deficit. It is worth emphasizing once again that in as much as the idealist subject itself disintegrates and redefines itself as the unique consumer profile, it is unlikely that this newly defined subject will experience any restriction. Strictly speaking, it does not have any dimension or potential which cannot be fully developed within the existing framework of the capitalist consumer economy.

Self-determination and self-development turn into the free development of the consumer-ego. The officially sanctioned demand that every individual ought to be able to realize its fullest *potential* essentially refers to the potential of income generation and spending power. Accordingly, the ideal of *equality* adopts the meaning of *equal opportunities* within the consumer economy. Equality thus no longer demands equal validity and appreciation of values and rationalities beyond the capitalist consumer system. At the same time, it emancipates itself from old imperatives of material redistribution and their outdated implications of self-restriction and social responsibility.

In this context, it is finally worth looking at the reflexive redefinition of the concept of *participation*. The aim of the participatory revolution in the 1960s and 1970s had been to counterbalance and control the logic of the dominant capitalist system. The struggle for individual participation in power, influence and responsibility was a struggle for plurality, diversity and individual autonomy. It was a struggle against the universalization of the economic system and for the project of a genuinely inclusive societal order. Participation implied the involvement of the individual as the carrier and agent of a genuinely rational, that is, human-centred, rationality that was meant to correct the irrational, perverted and alienating logic of the system. This involvement obviously could not mean to replace the restricted and restrictive logic of the system by an equally restricted logic of private interests and personal preferences. On the contrary, participation implied transcending individual self-interest, promoting the project of the common good and taking responsibility for the reinstatement of the genuinely inclusive society against the exclusive dynamics of a particular system.

In contrast to this, participation in the contemporary framework means, first and foremost, the realization of self-interest. Participation has changed its meaning from a generative process in which something is being created and shaped, and responsibility taken, into passive merely consumptive participation which restricts responsibility to the personal consequences of right or wrong choices. Participation does no longer imply making a contribution towards the realization of a common project and of the common good. Given the non-availability of such goals it has adopted the meaning of *making oneself heard* and *being listened to*. The primary objective of participation is thus to make sure that one does not miss out, to secure one's share of the wealth and the opportunities provided by the system. The rhetoric of the service society constantly reconfirms the belief that the individual represents the centre and goal of society. Accordingly, there can be no goals and agendas going beyond the vision, interest and capacities of the individual. Participation thus loses its important dimension of responsibility for, its reference to, the larger project and community. Its meaning is rapidly contracting into self-interested, self-orientated and self-responsible participation in the capitalist labour and consumer market.

Whilst the capitalist market economy gradually developed into an all-embracing global system, freedom, self-determination, autonomy, equality, identity, participation, and so on, have thus all been redefined in such a way that they no longer refer beyond this system. Whilst liberal politics continues to promote the liberalization of the market and to present market instruments as the ultimate solution to the existing societal problems, the theoretical possibility of any alternative to the universal logic of the market has virtually dissolved. The particular characteristic of this situation is that alternatives to the established system are not only no longer available, but that – because of the conceptual redefinitions – there is actually no demand for such alternatives. More than that: the colonization of alternative systems by the economic system, the universalization of the economic principle, is so complete that alternatives are actually no longer conceivable. After the redefinition of the constitutive elements of the project of modernity, there are no conceptual tools available which would allow even to think the alternative. Hence the experience of any deficit or of alienation in the Marxist sense can no longer emerge, and a desire for radical change and reconciliation does not exist. This is reflected in the absence of any lasting and co-ordinated protest movements and of genuinely alternative political parties and agendas.

The old emancipatory struggle against the colonizing forces of the capitalist economic system and for freedom and equality in the traditional sense has become outdated and anachronistic – to such an extent that these very deliberations about the colonizing and exclusive character of the capitalist system appear like the strangely retarded echo of intellectual debates from a bygone age. The struggle against the colonization of the other, against forced inclusion and for freedom, equality and autonomy, has been replaced by the struggle against exclusion from, and for inclusion into, the dominant (universalized) economic system. It is because not just the constitutive elements of inclusionality but inclusionality itself have adopted a radically different meaning that inclusionality can be retained as a goal and value, even though exclusionality has firmly established itself as a major structural principle of late modern societies. This conspicuous simultaneity of the rhetoric of inclusionality and the reality of exclusionality may be referred to as the contemporary *politics of simulation*. This politics of simulation and its eco-political implications will be further explored in the remainder of this chapter.

Simulative politics

The analysis so far has revealed how the concept of inclusionality has changed its meaning: (a) from inclusion of the individual *qua* representative of transcendental reason into inclusion of the individual *qua* subject of empirical interests and personal preferences (*inclusion of what?*); (b) from inclusion into the all-embracing rational context into inclusion into the global capitalist market economy (*inclusion into what?*); (c) from inclusion for the sake of the full realization of humanity's rational potential towards inclusion for the sake of the realization of the individual's full consumptive potential (*inclusion for what reason?*). The ideal of inclusionality has changed its meaning from an antidote to the universalization of any particular logic or rationality (at the expense of others) to a drug that stabilizes and reproduces the capitalist economic system and constantly reconfirms the exclusive validity of the economic code. The aim of reconstructing and conceptualizing this process of reflexive redefinition had been to explain why the ecologist appeals for a radical change in the way we think and for societal restructuring along the lines of a *participatory philosophy of inclusionality* necessarily remain futile. The explanation that had been suggested at the beginning of the Introduction is that late modern society – despite its continued ostentatious emphasis on social inclusion – has essentially become an exclusive society. This argument was developed proceeding from the observation that the principle of inclusionality once established itself as a moral and pragmatic imperative not because inclusionality was a value in itself, but because it seemed to serve the ideals of freedom, autonomy and the full development of the human potential. All of these had been defined with regard to the species rather than the individual and could therefore only be realized inclusively, that is, at the level of society – ultimately the world society – but not at the level of the individual. Following the redefinition of freedom, identity, self-determination and so on, however, inclusionality in the traditional idealist sense is not only no longer functional. It is actually counter-productive. Identity, autonomy, self-realization, self-expression, and so on, in the contemporary sense are much more efficiently and effectively promoted on the basis of the principle of exclusionality.

The principle of exclusionality can be said to have established itself as the major structural principle of late modern society, firstly in as much as, and because, the individual has become the ultimate point of reference, with the constitutive parameters of the modernist project being redefined with reference to the empirical individual rather than, as in the idealist tradition, to the transcendental rational subject. Individuality, individual identity always and necessarily relies on differentiation, on the erection of borders, on the exclusion of the Other. Secondly, the principle of exclusionality has been established as a fundamental structural principle of late modern society because, and in as much as, the economic system and the logic of the market have invaded (colonized) all other systems and marginalized all other ways of thinking. The system of the capitalist consumer economy has become global and all-embracing, suppressing and excluding all other systems. The decline of ecologist thought, which had endeavoured to establish itself as an independent logic and rationality, may be regarded as one of the latest casualties in this process. Under the banner of *ecological modernization*, ecological thought has, since the late 1980s, been converted to the beliefs of resource efficiency and managerialism, and has been fully synchronized with economic thought (Blühdorn, 2000a, 2001; Spaargaren, Mol and Buttel, 2000; Mol and Sonnenfeld, 2000; Weizsäcker et al., 1997; Weizsäcker, 1999).

The dominance of the economic system, the permeation of all social relations and activities by the code of payment and profitability, implies that the main dimension of individualization and the main pattern of identity formation will have to be economic, that is, the accumulation of spending power and material wealth. In a context which is predominantly governed by the principle of material wealth and consumption, material differentiation, that is, material inequality becomes the only, or at least the major, way in which individuality and identity can be established and expressed. To the extent that individual identity and material inequality have thus become inherently linked, it is justified to say that contemporary society is not only inherently exclusive but does not – despite its rhetoric and policies of social inclusion – have any ambition to become inclusive. The particular combination of the strong emphasis on individuality and personal identity on the one hand and contemporary society's one-dimensionality on the other, material differentiation, inequality and exclusion, are indispensable. There is no comparable means of identity construction and expression.

The exclusive character of late modern society however, is obscured – in order to avoid too much emphasis on conspiracy theories it may also be appropriate to say: disguises itself – by the fact that the issue of inclusion and exclusion remains in the centre of interest and public debate. Protest movements against economic globalization, newly emerging ethnic conflicts both within the rich industrialized countries as well as in the less developed world, new nationalisms and other fundamentalisms, or the huge problems of trafficking and migration all provide evidence that inclusion and exclusion still mark the crucial political cleavage. What is at stake in these conflicts, however, is no longer inclusionality in the idealist sense, that is, inclusion without exclusion, inclusion into the all-embracing system that leaves nothing outside. Instead, the focus is on redrawing the existing lines of exclusion in order to secure inclusion for oneself. As was indicated above, exclusion as such urgently has to be retained not just because individual self-construction, self-experience and self-expression depend on differentiation, inequality and exclusion, but in addition to this, because it is well known that the universalization of the desired wealth and standards of life is physically impossible. The dematerialization and virtualization of significant parts of the economy have expanded, but by no means removed, the limits to growth. Although late modern society prefers to systematically suppress any awareness and genuine debate of these limits, there is no doubt at all that the standards of material wealth in the industrial countries can only be retained, and the implicit and indispensable principle of growth sustained, if they remain restricted to a privileged minority. Significant parts of the global population as well as within particular national societies necessarily have to remain excluded. Not only in cultural terms (identity formation), but also in physical terms, social deprivation, marginalization and exclusion have thus become a necessary condition of late modern society.

In this context of structurally indispensable exclusion, the rhetoric and the policies of social inclusion serve to stabilize rather than overcome the system of exclusion. Just like the discourse and policies of *ecological modernization* and *sustainability* function to simulate the possibility and desirability of environmental justice and integrity (Blühdorn, 2000a, 2001) the discourse and policies of social inclusion simulate the validity of the old ideal of inclusionality whilst they factually promote the principle of exclusionality. Such policies are designed to reduce the costs of social exclusion, that is, public expenditure on social benefits and the costs of providing internal and external security (policing, border control). They are meant to reduce the tax burden to be carried by the middle and upper social strata and to facilitate their accumulation of wealth. Furthermore, such policies are intended to increase the spending power of the less well off and turn them into socially pacified and economically profitable consumers. Policies of social inclusion thus function to stabilize and reproduce, but never to overcome the underlying principle of exclusion. Genuine material equality however, or a genuinely equal distribution of political influence and control, is neither intended nor desirable.

This argument may be further developed by taking into consideration that the desire for inclusionality did not just emerge from the modernist belief that the human potential that has to be realized is a social potential rather than the potential of any individual. Of equal significance for the emergence of the ideal of inclusionality was, arguably, the *fear of the Other*, the threat that emanates from the unknown. The project of inclusionality was driven by the desire for controllability, calculability, predictability of the unknown, the unforeseeable, the threat. But calculability and predictability require structural uniformity. The desire for inclusionality therefore enforced uniformity and led to the suppression of the Other, including the suppression of nature and the environment. Up to the present, policies of social inclusion are arguably driven by the threat that emanates from the socially excluded. What is at stake is clearly the efficient management of social inequality and exclusion and not the desire to genuinely share social wealth and opportunities, and grant the hitherto excluded the right to develop *their* potentials rather than the potential that mainstream society seeks to impose on them.

The discourse and policies of inclusionality thus amount to a politics of simulation. Yet it would be a fatal error to misunderstand *simulation* in this context simply as the conspiratorial deception of the underprivileged by certain social elites. Although this may well be one important dimension, the politics of simulation simulates in a much more comprehensive sense the ongoing validity of the modernist project for society as a whole. Social elites who might at first sight appear as the originators and agents of consciously constructed and strategically deployed simulations, are therefore at the same time themselves the objects and victims of this deception. Given that the politics of simulation involves and affects society as a whole, that it is a societal strategy of societal self-deception, the particular difficulty of any analysis of *simulative politics* is that there is, actually, no perspective from which this practice could be revealed as a conscious strategy. It is difficult to identify any subject to which it could appear as a deception and which could criticize this practice as a simulation.

What the societal practice of simulative politics simulates is not only the validity of the ideal of social inclusion, or appreciation for the value of environmental inclusion (sustainability), but reaching down to a far more fundamental level, simulative politics functions to stabilize and reproduce the belief in the existence and relevance of the individual. Beyond the pacification and instrumentalization of the underprivileged for the benefit of social elites, the contemporary politics of inclusionality simulates that contemporary society and its function systems are really centred around the needs and interests of individuals, and the belief that material accumulation and exclusion do really establish individuality and self-identity. Thus, the politics of simulation serves not just to deceive the underprivileged, but, as was indicated in the previous section, it serves to simulate the relevance and centrality of human individuals in general (even the so-called rich and powerful). In a context where the modernist and humanist essentials seem to be marginalized by post-modernist and post-humanist conditions this is an important strategy of reassurance.

The rhetoric of social inclusion, reinforced by the discourse of individualization and the talk of the service society, reconfirms the belief that more than ever the individual and its interests and needs represent the centre of society, its ultimate point of reference and the legitimating ground for all societal progress and development. In a life world sense this is actually the case, for the pluralization, flexibilization and temporalization of norms and values specifying the underlying code of profitability and accumulation – Luhmann (1995) would say: specifying the *programme* that makes the self-referential *code* meaningful – have left self-interest and personal desires as the main yardstick and goal. In a context of desolidarization, competition and exclusion, responsibility has adopted the primary meaning of self-responsibility. However, as the analysis in the previous section revealed, individual identity has adopted the shape of the individual consumer profile with the individual-centred life world disintegrating into an amorphous range of product choices. In other words, the individual has very little or no substance and identity beyond the market. Against this background, it becomes increasingly important for contemporary society to reassure itself that there is something beyond the economic system that can infuse the economic code with meaning. The contemporary celebration of individuality, the constantly reiterated assertion that contemporary society is a *service* society with all its institutions being *service* providers to the individual to whom all social institutions are accountable, function to ensure exactly this: they simulate the existence, relevance and centrality of the individual; they simulate duality in order to disguise self-referentiality. This simulation invisibilizes the fact that the centre of the comprehensively monetarized and commercialized service society is, of course, not the autonomous self but the valued customer. They invisibilize the postmodernist and post-humanist self-referentiality of the globalized economic system.

In a wider sense, the politics of inclusionality that simulates the existence and relevance of the individual, may also be said to simulate the possibility and reality of politics. In a context where participation has adopted the primary meaning of passive and consumptive participation in the market, the politics of inclusionality generates the impression that the individual is actively in the driving seat and in control of societal development. Factually, there is neither much interest in taking control of societal progress, nor is there much confidence that politics really matters. The project of giving societal progress a purpose and direction is faced with the dual dilemma of the normative emptiness and disorientation of politics on the one hand, and the disempowerment of politics by the necessities of the global market on the other. Nevertheless, the belief in politics (beyond the politics of self-interest) is indispensable, firstly, because – despite all contemporary fatalism – politics still does retain a certain influence on the economic system, and secondly because politics, irrespective of the real motor of societal development, has to fulfil the important function of providing it with legitimacy. In this sense, the contemporary politics of inclusionality does not only simulate the validity of ideals like social and ecological inclusionality, but it is also the simulation of the possibility and reality of politics. In a context where the potential of politics has become questionable, simulative politics is not just the politics of simulated inclusionality, but is at the same time the simulation of politics. At the risk of overstating the point it might be said once again, that what is being simulated is the dualism that is essential for the project of modernity, and whose collapse would imply the transition from *late-modern* to genuinely *post-modern* and *post-humanist* society. Despite all this, however, it would be wrong to denounce the discourse of inclusionality, that is, of equality and participation, as pure rhetoric without any genuine commitment and empirical substance. The struggle for social inclusion is genuine and bears empirical results. Equality and participation are being implemented – yet what these concepts signify has radically changed. The politics of inclusionality thus facilitates the emergence of post-modern society by simulating the continuation of modernity. The ecologist talk of an immersive philosophy of inclusionality invariably contributes to this simulation of modernity and distracts attention from the new post-ecologist ways in which contemporary society formulates and processes its ecological issues.

Conclusion: towards an environmental research agenda

The aim of this comprehensive chapter has been to bring new momentum into an ecological debate that appears to be deadlocked into forever repeating its futile appeals for an *immersive philosophy of inclusionality*. The objective was to explain why ecologists, with their demand for 'a radical reorientation in the way we think' (Rayner) necessarily have to remain callers in the desert. By exploring the contemporary meaning and significance of *inclusionality* and by reinterpreting this term within the conceptual framework of *simulative politics*, this chapter has aimed to demonstrate how the principle of *exclusionality* has firmly established itself as a main structural principle of late-modern society. The shift from the principle of inclusionality to the principle of exclusionality is at the centre of the shift from *ecologist* politics to *post-ecologist* politics (Blühdorn, 2000b). Within the *post-ecologist constellation*, there is neither any space for *inclusionality* nor for an *immersive philosophy of environmental relationships*. If ecologism could be described as the battle against the oppression and exclusion of the societal and extra-societal Other, post-ecologism may be referred to as the battle for self-inclusion into a system that is known and accepted to be inherently exclusive.

The chapter has demonstrated that ecologists are, of course, right in regarding the issue of inclusion and exclusion as the core issue in the eco-political debate. But they are fundamentally wrong in assuming that the principle of inclusionality has any genuine appeal to contemporary society and thus any chance of providing a basis for a restructuring of late-modern society. Ecologists may be said to have fallen prey to – and be helping to reproduce – the politics of simulation. Hence they urgently need to review the sociological and normative foundations of their arguments. Correctly, Rayner is pointing towards an 'attitude problem'. He is right in saying that 'the first thing is to admit that it really is a problem' and 'to know just how big a problem it is'. And indeed, the next step must then be to 'find an imaginative way out of the trap'. Exactly this has to be the agenda – but not for the reorientation of society, but for the reorientation of ecological thought and environmental sociology.

The *radical reorientation* which ecologists have always demanded has two dimensions, firstly a reorientation in the way we think, and secondly a reorganization of societal structures and practices. Late-modern society cannot realistically be expected to perform either of these. It cannot be expected to rethink, because it does not think at all – it does not have subject status. It cannot be expected to restructure, because it does not have any strategic centre that is capable of planning, co-ordinating and implementing such a reorganization. The development of late-modern society is determined by a combination of systemic necessities and unreflected egoisms which are – after all – not integrated by any invisible hand, world spirit, cunning of reason, historic necessity, or whatever other principle modernist thinkers from Smith to Marx had tried to identify. Late-modern society lacks the normative foundations, the political will, the political actor and all other preconditions required for any reorientation in the ecologist sense. Against this background, any realistic expectations for a reorientation must be addressed to ecological thought itself. Such a reorientation can be expected, in particular, from environmental sociology. It can be performed if environmental sociology emancipates itself from established ecological dogmas and imperatives of ecological correctness. For environmental sociology exploring the changing discourse of inclusionality and the new relationship between inclusionality and exclusionality represents a major challenge. Its particular task is not to try and solve ecological problems or to preach the big ecological U-turn, but to provide a sociological analysis and assessment of the post-ecologist constellation. Its task is *sociological enlightenment* in Niklas Luhmann's sense (Blühdorn, 2000c) rather than *ecological enlightenment* in the sense of Ulrich Beck (Beck, 1995).

In conclusion to these suggestions for a new environmental research agenda it needs to be emphasized once again that the objective must not be to declare the modernist and ecologist goals and ideals normatively wrong. Irrespective of the normative validity these ideals may have or lack, environmental sociology – in contrast to ecologist philosophy – restricts itself to exploring their appropriateness for late-modern society, that is, exploring the extent to which late-modern society can relate and respond to them. Thus, the intention of this chapter is not to discard the normative beliefs underlying Alan Rayner's argument. On the contrary, the deliberations of this chapter are fuelled by the belief that the old ecologist values ought to be honoured. However, if these values can be rescued at all, then the only way of contributing to this is that environmental sociology ruthlessly tears down the green veil that ecologists unwittingly help to weave and that late-modern society likes to wear in order to hide the injuries providing evidence of its transition to the age of *post-ecologist politics*.

References

Bauman, Z. (2000) *Liquid Modernity*. Cambridge: Polity.
Beck, U. (1995) *Ecological Enlightenment. Essays on the Politics of the Risk Society*. New Jersey: Humanities Press International.

Beck, U. (1997) *The Reinvention of Politics. Rethinking Modernity in the Global Social Order*. Cambridge: Polity.
Beck, U., Giddens, A. & Lash, S. (1994) *Reflexive Modernization. Politics, Tradition and Aesthetics in the Modern Social Order*. Cambridge: Polity.
Blühdorn, I. (2000a) 'Ecological modernisation and post-ecologist politics'. In Spaargaren et al. (eds), pp. 209–28.
Blühdorn, I. (2000b) *Post-Ecologist Politics. Social Theory and the Abdication of the Ecologist Paradigm*. London/New York: Routledge.
Blühdorn, I. (2000c) 'An offer one might prefer to refuse. The systems theoretical legacy of Niklas Luhmann'. In *European Journal of Social Theory*, 3(3), 339–54.
Blühdorn, I. (2001) 'Reflexivity and self-referentiality: on the normative foundations of ecological communication'. In Grant et al. (eds), pp. 181–201.
Grant, C. & McLaughlin, D. (eds) (2001) *Language, Meaning, Social Construction. Interdisciplinary Studies*. Amsterdam: Rodopi.
Habermas, J. (1984) *The Theory of Communicative Action. Volume 1: Reason and the Rationalisation of Society* (translated by Thomas McCarthy). Boston: Beacon Press.
Habermas, J. (1987a) *The Theory of Communicative Action. Volume 2: Lifeworld and System: A Critique of Functionalist Reason* (translated by Thomas McCarthy). Cambridge: Polity.
Habermas, J. (1987b) *The Philosophical Discourse of Modernity. Twelve Lectures* (translated by Frederick Lawrence). Cambridge: Polity.
Kant, I. (1784) *Idea of a Universal History with a Cosmopolitan Purpose*. In Hans Reiss (ed.) (1970) *Kant. Political Writings* (translated by H.B. Nisbet), pp. 41–53. Cambridge University Press.
Luhmann, N. (1995) *Social Systems* (translated by Rhodes Barrett). Berlin/New York: de Gruyter.
Mol, A. & Sonnenfeld, D. (eds) (2000) *Ecological Modernisation Around the World. Perspectives and Critical Debates*. London: Frank Cass.
Spaargaren, G., Mol, A. & Buttel, F. (eds) (2000) *Environment and Global Modernity*. London/Thousand Oaks/New Delhi: Sage.
Weizsäcker, E.U.v., Lovins, A. & Lovins, H. (1997) *Faktor Vier. Doppelter Wohlstand, halbierter Naturverbrauch*, 11th revised edition. Munich.
Weizsäcker, E.U.v. (1999) *Das Jahrhundert der Umwelt. Vision: Öko-effizient leben und arbeiten*. Frankfurt: Campus.
Young, J. (1999) *The Exclusive Society*. London/Thousand Oaks/New Delhi: Sage.

Part II

Environmental Education

3

Environmental Education and the Arts–Science Divide: The Case for a Disciplined Environmental Literacy

Andrew Stables

My purpose in this chapter is to draw attention to the limitations of the epistemologies we have inherited from Renaissance humanism and Enlightenment rationality with regard to the possibility of an education for environmental sustainability. I shall suggest that more than cross-disciplinarity, and more than interdisciplinarity as conventionally defined (that is, as mixture of arts and science) is required, yet that we are constrained by inherited traditions of thought to do no more than move slowly and tentatively in this direction. First, I suggest that we must attempt to transcend these traditions through a greater emphasis on their cultural histories, so that we can appreciate both where they have come from and where they seem to be leading us; secondly, I argue for a more disciplined approach to the cultivation of forms of what, for want of a better phrase, I shall term 'environmental literacy'.

An influential tripartite model of the school and college curriculum, proposed originally by Malcolm Skilbeck (Skilbeck, 1976) and adopted by Denis Lawton (Lawton, 1989) discriminates between a Classical Humanist, a Progressivist and a Reconstructionist view of the curriculum. According to this model, the Classical Humanist approach has knowledge divided into fairly stable subject disciplines, the essential aspects of which should be transmitted through the educational process. Education is thus a process of cultural assimilation via cultural heritage. The Progressivist view places the readiness and interests of the learner at the centre, and resulted in the child-centred approaches particularly dominant in primary education in England and Wales. It is an approach strongly influenced by both Piagetian Constructivism and Wordsworthian Romanticism, a view which valorizes the child's self-expression while stressing the importance of cognitive readiness for the next level of cognitive challenge. The Reconstructionist approach sees education as the building blocks of a new and better society (as defined, of course, by its dominant adults). It views education principally as an agent of social change, an instrument of social justice.

It is not the intention here to critique Skilbeck's model, but rather to note that it acknowledges three principal foci for education: epistemological tradition (the Classical Humanist model), the personality of the learner (the Progressivist model) and the nature of human society (the Reconstructionist model). While a focus on each of these areas may well lead to a consideration of human relations with the non-human world under certain conditions (such as when threatened by pollution, or moved by sentiment at the threat to an endangered species), the notion of *humanity in the non-human world*, as inextricably related to and part of it, is secondary to each.

Effectively, each of Skilbeck's approaches, not merely the first, is humanistic in its orientation. Post-Renaissance humanism has left us with both academic and political traditions and structures that are not primarily concerned with ecology in its broadest sense. Our thoughts and our actions are still principally orientated to what seems an increasingly narrow humanistic ideal.

Politically, the dominant world views associated with capitalism and socialism are each equally concerned with the accretion and distribution of material wealth for human beings. Effectively, the differences relate only to the perceived purposes for the accretion of that wealth (primarily for its own sake, or to meet defined human needs) and to the channels of distribution (through family and inheritance, or across societies, through collective agreement or central diktat). Environmentalists who are still locked into arguments over the supremacy of one of these approaches over the other are arguably missing the point that neither was developed to achieve what one supposes to be their aspirations. Both capitalism and socialism rely on environmental sustainability, insofar as each relies on an apparently limitless supply of natural resources to attain its material ends, but for each, environmental sustainability is a prerequisite, and not an aim. In recent years, some environmentalist critical theorists have begun to acknowledge the limitations of the marxist models that rest on the premise that social justice is synonymous with sustainability. The eco-critic Greg Garrard, for example, has written an interesting piece on the notion of 'radical pastoral' in the journal *Studies in Romanticism*, in which he begins to tackle some of these very issues in the literary-critical tradition: a tradition in which writers over the last decade have felt the need, for example, to attempt to reinstate Wordsworth as a nature poet, such was the strength of an anthropocentric new historicism grounded in a late marxist critical orthodoxy (Garrard, 1996).

Similarly, our disciplinary traditions do not have ecological or environmentalist ends. Man's (*sic*) increasing post-Renaissance faith in his ability to create new and better models to describe his universe, based on his own observations and experiences, has ironically resulted in what has been, and might still be, crudely described as the arts–science divide. While the Enlightenment posited models of natural behaviour (that is, the behaviour of nature) that, while developed through human observation, seemed to take no account of human behaviour and relationships, our understanding of 'nature' itself split in two. On the one hand, we had developed through logic and observation a model of a mechanistic universe; on the other, the desire to express and understand the felt and the immediate resulted in the inevitable reaction of Rousseau and Romanticism against the new science. The self as subject was compelled to reassert, both individually and (perhaps to a lesser extent) collectively, its irrationality, or its more-than-rationality, under the hegemony of Cartesian dualism. The word 'nature' itself, which in Shakespeare's day referred to the living soul of all living things, to that into which we are born, became on the one hand the disembodied and mathematical totality of physical forces, and on the other the increasingly solipsistic experience of the Romantic ego. As T.S. Eliot put it, though he was ostensibly referring only to literary style, 'In the seventeenth century a dissociation of sensibility set in from which we have never fully recovered' (Eliot, 1964). It seemed as though thought and feeling were no longer one.

Whatever the protestations of the most gifted scientists or the most idealistic educationalists, few would dispute that in the mainstream educational experience of the many, there is still an arts–science divide. For example, as Habermas (1987), observed, positivist science finds its application in an instrumental rationality which is good at getting things done but not good at dealing with the issues of what should be done, or even the consequences of what it has done. Interpretive and critical-emancipatory knowledge might take their cues from positivist science, but bear little resemblance to scientific knowledge. The arts, on the other hand, tend to be seen as having the power to move, and therefore the capacity to inform and sensitize our ethics, but are given little credibility as prescriptions for action. Each of these assumptions is, of course, entirely historical and neither is immutable; when science was cosmology, technology was art. (The Greek word for 'poet' means 'maker'.) Nevertheless, we live with what we are, and simple rejection of science and technology, or even the arts, is entirely counter-productive.

Support for the assertion that neither positivist science nor humanist art exists centrally to promote ecological aims comes from a critique of the role of the social sciences in fulfilling this function. As 'halfway between' science and arts, a good disciplinary home for environmental study would seem to lie within the social sciences. However, social science is highly problematic in this respect. Alan Holland (1994), for example, has written convincingly on the problems of the application of the concept of 'natural capital' in a volume edited by Robin Attfield and Andrew Belsey under the title *Philosophy and the Natural Environment*. The theoretical base of classical sociology seems limited in its power to conceptualize environmental and ecological issues beyond certain aspects of environmental policy. Issues of class, gender and access to resources, for example, contribute to the environmental debate but do not deal with it any more fully than positivist science or interpretive art.

So where does this lead us? What we cannot achieve is the waking enactment of a simple dream of holism. As I shall later allude to the textuality of environment, and as 'text' is etymologically related to 'texture' or 'weaving', it should be stressed that we cannot undo the yarns that have woven us. Thus, for example, primary school teachers who believe that subjects do not matter run the risk of disempowering the children in their charge by failing to give them control over the discourses which constitute knowledge, while environmentalists who preach a kind of New Age holism without reference to its epistemological and ontological roots may be equally naive.

We can, however, understand some of the yarns and forms of weaving that bind us, and we can continue the weaving in our own ways. We can both understand where history and science have led us, and how they have led us there, and modify our practices. Indeed, we shall. Subject disciplines *are* discourses, and the point has been made by many over the last half century that we do not simply *use* language, but *change* it, slowly and collectively, through our usage.

I now come, therefore, to two suggestions as to how the educational process can begin to change in a way which may, in the long term, lead to some more fruitful coexistence of the arts and the sciences in terms of meeting aspirations beyond the narrowly humanistic.

The first is that we should do what we can as educators to make our students understand how knowledge has been assimilated and used in response to changing perceptions of need over the centuries. An induction into the history and philosophy of science should run parallel with, and not trail far behind, an induction into the current assumptions and practices of science, and the latter should be treated less uncritically than they still often are, at least at the school/college level. In the arts and humanities, the thorny issues of ideology and hegemony should be tackled head-on and not swept under the carpet of superficially value-free teaching. The historical interrelatedness of issues that we now treat as subject-distinct could be much more fully explored. For example, to what extent can Wordsworth be held responsible for the creation of the English Lake District as we now know it? (Prior to the close of the eighteenth century, few took any interest in such barren highlands.) What are the implications for the consideration of specific conservation issues within the English National Parks generally that flow from such a debate? How many local councils are basing their environmental protection schemes on pastoral visions of eighteenth-century parkland, and how do the biodiversity requirements flowing from Agenda 21 interact with this nostalgia? In terms of the urban landscape, how has our industrial heritage altered conceptions of, and assumptions about, the city, and how can this be taken into account in attempts to respond to southern English people's current common desire to leave in order to live an essentially suburban life among green fields, thus adding to traffic problems and the like? It can be argued that informed consideration of issues such as these requires an interdisciplinarity born of specialist historical, scientific and cultural knowledge. Certainly, the debates can only be understood in terms of their histories.

Much of my own inspiration in thinking along these lines came from reading Simon Schama's *Landscape and Memory* (1995). There is so much to be done through (for example) a cultural history of the landscape, in terms not only of illuminating but also of redefining, environmental issues: much more, in fact, than Schama himself acknowledges.

Secondly, I propose that the nearest we can get at present to a term which acts as an umbrella for the kind of multidisciplinary enquiry I am promoting is 'environmental literacy'. However, hitherto this term has often been used uncritically and has not been examined with the rigor necessary to enable it to exploit its educational potential. Where 'environmental literacy' has been used hitherto in the environmental education literature, any working definitions that have been evoked have not been derived directly from a systematic engagement with literacy debates within language and literature studies, where the rigorous examinations of 'literacy' have generally taken place.

Examples of previous engagements with the concept of environmental literacy include the following. As part of the American work on 'standards', Roth (1992) has provided a framework for environmental literacy with relation to knowledge, affect, skills and behaviour at three levels of competence (nominal, functional and operational). For UNESCO, Marcinkowski (1991) provides a set of nine statements which amount to what environmental literacy might be taken to be, relating to knowledge, understanding, attitudes and active involvement. In Scotland, curriculum planners have included environmental literacy as one of the four goals of 'environmental citizenship' defining it in terms of 'knowledge and understanding of the components of the system' (Scottish Office, 1993). While each of these definitions of environmental literacy might have its practical uses, none is overtly grounded in the primary academic debate about the nature of literacy.

I have argued the case for the importation of language and literature studies into the environmental debate in a series of earlier papers (Stables, 1993, 1996, 1997, 1998), and the arguments would take too long to rehearse fully here. At its simplest, all that is argued for is a further minor twist in the linguistic turn in twentieth century philosophy which has raised language from that which simply mirrors or carries ideas to that which, to a degree at least, constitutes them: a shift from representation to re-presentation. Similarly, and by extension, 'language' in its broadest sense has come to be that which is not merely embodied in words and sounds, but embraces semiotic decoding and encoding in all its forms. Additionally, of course, language and literature studies explore personal responses and meanings and not merely impersonal truths.

There are strong theoretical arguments for regarding our response to our surroundings as essentially semiotic. What things are is what they mean to us, in terms of our various 'languages' of science, myth and tradition. Whether or not this constitues a relativist position, it is not a nihilistic relativism. What things mean to us is what they are. The world is the sense we make of it: but not just as individuals, for there are dominant ways of looking at the world, and they matter most. Environmental literacy is – should be – about understanding and contributing to those ways of looking at and interacting with (decoding and encoding) the world.

The explanations of environmental literacy below come from the debate about print literacy, because it is within this dominant tradition that the issues have been most thoroughly and convincingly explored. (To discuss literacy from an intrinsically different perspective really would be a lapse into nihilistic relativism.) Within that debate, there is sometimes used a tripartite division of literacy skills, which it is useful to apply to the concept of environmental literacy: a division into functional literacy, cultural literacy and critical literacy. I shall now say a little about each of these with specific reference to the environment. Some of the following material was published first in an article in the journal *Environmental Education Research*.

In the simplest terms, functional literacy refers to the ability to understand at a superficial level and to act on instructions. It implies the ability to read but not to reflect. Functional literacy is not, however, just a matter of knowing what words mean, but of being able to *find out* what they mean in the context of whole sentences by the use of phonic and contextual cues. Functional print literacy also involves being able to read words referring to commonplace abstractions (beauty, goodness, fear, and so on). It involves literal comprehension.

Functional *environmental* literacy must, therefore, refer not only to the ability to remember what an oak tree is, but to recognize one; not only to recognize several trees within a given area, but to know whether they form part of a wood or an area of parkland. Functional environmental literacy must also involve the ability to ascertain, from contextual cues, what something half known is likely to be: for instance, to make an informed guess, using observation, at the types of woodland flower within a beech copse overlying chalk rather than an oak wood on more acid soil. Additionally, functional environmental literacy must encompass understanding of basic biological, chemical and physical processes, from food chains to global warming. Functional literacy is not, therefore, a mere prerequisite to more advanced forms of literacy, but involves a series of complex skills and an accumulation of knowledge which has unlimited capacity for growth. Arguably, much science education in schools focuses chiefly on what is defined here as functional environmental literacy, whether or not this entirely reflects its intentions. Certainly, its role in environmental education should not be underestimated.

Both cultural and critical literacy are impossible without functional literacy. Just as the ability to decode print is a prerequisite to the development of deeper levels of comprehension of the passage to be read, so is knowledge of the natural world a condition of the development of awareness of environmental issues and of the ability to take effective action.

Also, functional environmental literacy is not enough because it does not, of itself, engage the learner (though many learners may already be highly motivated), and it does not, in this case, engage either with the crucial notion of what the environment *means*, either to others or to the learner. In terms of environmental literacy, we must acknowledge the importance of the functional but place it alongside the cultural, and see both as conditions of the critical, as only critical environmental literacy can facilitate effective environmental action.

To sum up, 'functional environmental literacy' might be a term used to coalesce efforts at understanding at a 'literal' level how the natural world works and how it interacts with the human world. The content-loaded study of subjects such as biology and chemistry in many schools seems to function almost entirely at this level (despite the wishes of some educationalists). Yet, despite its limitations, this kind of knowledge is of central importance. Nobody can make a real contribution to the ecological debate without some level of what many loosely refer to as 'scientific knowledge and understanding'.

Cultural literacy (following E.D. Hirsch's (1987) controversial work in the United States) refers to the ability to understand the social and cultural significance of things: that bonfire night is not just about having a fireworks party, for instance; that historically, the solstices had roles in patterning human lives and beliefs, the vestiges of which are still evident in contemporary practices, such as the celebration of Christmas. Cultural literacy refers to the ability to understand the significance that societies attach to cultural icons. Such icons include living natural objects: national parks; the Californian redwood; the English oak. An increased cultural environmental literacy would be gained by a reading of Schama's *Landscape and Memory*, in which the author discusses a series of landscapes of rich significance to contemporary societies, including part of the Eastern European forest, the English Greenwood and the Californian redwoods, in terms of cultural history with respect to the ways in which these landscapes have been viewed, used and reshaped over a millennium.

On one level, a degree of cultural environmental literacy merely enables one to recognize the significance of natural images in human culture, along with some recognition of why and to whom they are significant: a red rose or the white dove of peace, for example. However, it also allows for an understanding of why the landscape itself is as it is, shaped not merely by climate, glaciation and topography, but also by arguments about enclosure, the need for timber and patterns of land ownership dating back many centuries. While functional environmental literacy develops knowledge of what natural things are, cultural environmental literacy enables us to explain why they are there when the causes are clearly not simply geological or climatic with no apparent human intervention.

Cultural literacy depends on a degree of acceptance of cultural hegemony: it links the learner with a dominant value system. The culturally literate individual in England will know what is implied by the term 'heart of oak', or understand the English Lake District as a kind of symbol of Wordsworthian Romanticism, even though these conceptions may be more associated with English 'high culture' than with popular culture, as well as having no scientific basis. Cultural literacy refers more to cultural heritage than to cultural analysis. The subtitle of Hirsch's book is 'What every American needs to know'.

Cultural environmental literacy would refer to an understanding of, for example, the social and cultural significance of the countryside in contemporary southern England. Without such a level of literacy, there are no grounds for environmental debate. Insofar as cultural literacy is empowering, however, it empowers by giving the learner access to socially powerful perspectives; cultural literacy alone does not enable the learner to act upon that knowledge, once acquired. Effective environmental action requires critical environmental literacy.

Critical environmental literacy is what many environmentalists are ultimately after. It is characterized by the ability to take action on environmental matters in the light of a good understanding of 'where the issues come from'. Without functional and cultural environmental literacy it is, however, either downright dangerous or simply impossible. The term 'critical literacy' is taken here to relate both to the ability of the individual to critique, in the liberal-humanist tradition, and to Habermas's more politically explicit notion of emancipatory knowledge: critical as in 'socially critical', leading to liberation through exposure and critique of underlying ideology. Critical literacy in textual studies implies the ability to understand the text on a deeper and more creative level: the ability to discuss the use of genre in context, to question the motives and ideology of the text, and to explore and develop personal (and broader social) response to it. Critical environmental literacy must then imply the power to develop an understanding of the factors that contribute to environmental change and to have a view on how to further or oppose that change in a way which can be translated into action. Critical environmental literacy involves the ability to explore questions such as: 'What does [a place or an issue] mean to me?'; 'What does it mean to us, or to others?'; 'What are the consequences of carrying on in this way [in relation to this place or this issue]?'; 'Should we act differently, and if so how, and what might be the consequences of that different action?'; 'How do we translate our values into effective action – and are our values themselves ready for change as a result of what we now know or feel?'.

As has been stressed above, critical literacy cannot be effectively developed without good levels of both functional and cultural literacy, though the latter are arguably pointless without the former. Critical environmental literacy relies on functional environmental literacy because both environmental debate and environmental action rely on information. Critical environmental literacy relies on cultural literacy not simply because environmental debate and action need to be grounded in an awareness of the norms and values of, say, national, ethnic and local cultures, but because influence on environmental change demands an understanding of the norms and values of the *dominant* culture.

In relation to these three forms of environmental literacy, it seems that curriculum planners in England and Wales have expressed ideals relating to the promotion of the critical but have not really addressed the issues involved in realizing them. A critical look at the (former) School Curriculum and Assessment Authority's non-statutory guidance on environmental education (1996) for example, reveals ambitious aims to do with young people as guardians of the world of the future, some of them expressed by Conservative government ministers, but focuses at the level of detail almost exclusively on low-level learning activities within school science and geography.

There is much more to it than the statement of lofty aims. It is by no means clear at this stage, notwithstanding the above arguments for environmental literacy, that an education for environmental sustainability is even possible: apart from anything else, there are many complex philosophical issues involved in the assumption. What does seem clear however, is that the traditions we have inherited, in politics and the academe, are not principally orientated to the attainment of any such sustainability. The problem is for us is to move beyond these traditions and the post-Renaissance humanism which bred them without being anti-humanist (within that broader tradition, there have been plenty of lessons in how to be anti-human). Any less anthropocentric framework for environmental education must acknowledge both that science and the arts as they stand are all that we have so far got, and that humanity is quite capable of transcending its own boundaries of thought – a little at a time. I propose that for the short term, a disciplined approach to the development of environmental literacies, in the terms I have begun to sketch out, may at least be a way of engaging multidisciplinarity without ending up, face down, in a cross-disciplinary or even interdisciplinary mush.

References

Eliot, T.S. (1964) *Homage to John Dryden: Three Essays on Poetry of the Seventeenth Century*. New York: Haskell House.

Garrard, G. (1996) 'Radical pastoral?' *Studies in Romanticism*, 35(3), 449–65.

Habermas, J. (1987) *Knowledge and Human Interests*. Cambridge: Polity Press.

Holland, A. (1994) 'Natural capital'. In R. Attfield & A. Belsey (eds), *Philosophy and the Natural Environment*, pp. 169–82. Cambridge: Cambridge University Press.

Hirsch, E.D. (1987) *Cultural Literacy: What Every American Needs to Know*. Boston: Houghton Mifflin.

Lawton, D. (1989) *Education, Culture and the National Curriculum*. London: Hodder & Stoughton.

Marcinkowski, T. (1991) 'The relationship between environmental literacy and responsible environmental behavior in environmental education'. In M. Maldague (ed.), *Methods and Techniques for Evaluating Environmental Education*. Paris: UNESCO.

Roth, C. (1992) *Environmental Literacy: its Roots, Evolution and Direction in the 1990s*. Ohio: Ohio State University.

Schama, S. (1995) *Landscape and Memory*. London: Harper Collins.

School Curriculum and Assessment Authority (UK) (1996) *Teaching Environmental Matters Through the National Curriculum*. London: HMSO.

Scottish Office (1993) *National Strategy for Environmental Education in Scotland*. Edinburgh: HMSO.

Skilbeck, M. (1976) 'Ideologies and values'. Unit 3 of Course E203, *Curriculum Design and Development*. Milton Keynes: Open University.

Stables, A. (1993) 'English and environmental education: the living nation in Macbeth'. *The Use of English*, 44, 218–25.

Stables, A. (1996) 'Reading the environment as text: literary theory and environmental education'. *Environmental Education Research*, 2(2), 189–95.

Stables, A. (1997) 'The landscape and "the death of the author"'. *Canadian Journal of Environmental Education*, 2, 104–13.

Stables, A. (1998) 'Environmental literacy: functional, cultural, critical. The case of the SCAA guidelines'. *Environmental Education Research*, 4(2), 155–64.

4
Education and Training for Sustainable Tourism: Problems, Possibilities and Cautious First Steps

Stephen Gough and William Scott

Tourism and ecotourism

Global tourism is the world's biggest and fastest growing industry (Filion et al., 1994). To many governments the expansion of tourism appears to be a very attractive method of achieving economic growth. Potential hard currency earnings for successful host countries are large. The expectation of such earnings is central to the development plans of a number of small states (Cater, 1995) and an important component in the strategies of many larger ones. Other advantageous aspects of tourism growth may include the creation of (often usefully decentralized) employment, enhanced tax revenues, a stimulus to conservation efforts, the attraction of inward foreign investment, and the creation of economic and recreational infrastructure for local use (Alderman, 1994; Pleumarom, 1994).

Ecotourism has been estimated to account for between 40 and 60 per cent of international tourism and at least 25 per cent of domestic tourism. Annual economic impacts from ecotourism may be as much as US$1.2 trillion (Filion et al., 1994). Doubts about these figures arise not only from problems of data collection and processing, but also from the lack of any universal definition of ecotourism. In particular, there is a tendency for definitions to widen as the perception grows among host governments and tour operators that an ecotourism label can aid marketing. From an academic perspective, the term has been defined simply in terms of what the tourist sees and does (Filion et al., 1994; Croall, 1995) and with additional references to the environmental and cultural impacts the tourist has (Tickell, 1993). The Ecotourism Society, based in Virginia, USA, includes in its definition a requirement that the well-being of local people be sustained (Johnson, 1998, p. 9; Roberts, 1998, p. 16).

Sustaining well-being is not necessarily the same thing as conserving culture, though the two could go hand in hand.

Ecotourism appears to be a highly segmented market, in which the following categories may be found: rough ecotourists, smooth ecotourists, specialist ecotourists, scientific tourists, cottage tourists, wildlife tourists, wilderness tourists, safari tourists, designer tourists, risk tourists, adventure tourists, alternative tourists, sensitive tourists and post-industrial tourists (Mowforth, 1993; Cater and Goodall, 1992). The important point, however, is that all such tourists are participants in a global, multi-billion dollar business which, to a greater or lesser extent, both shapes the environment and consumes it (Goodall, 1995). There is no such thing as zero-impact tourism (Lawrence, 1994).

Of course, attempts may be made to control the impacts tourism has, but even if the contribution made to global environmental change by tourism transport services is ignored, concepts such as 'tourism carrying capacity' are extremely difficult to operationalize. The attempt to satisfy the yearning of many who consider themselves ecotourists for pristine or traditional environments may be self-defeating and lead to progressive encroachment into remote areas, since once a place is easy to visit much of its appeal may be lost (Cater and Goodall, 1992; King, 1993). Ecotourist excursions of a few days duration or less are often mounted from bases in luxury hotels, and constitute a only a fraction of total holiday time. For all these reasons it may not, in the end, be very helpful to distinguish ecotourism from the mainstream. This is not to say that environmentally responsible tourism is impossible. However, very little is accomplished by establishing definitional criteria for 'ecotourism' if these are widely ignored, or if activities denied an ecotourism designation continue anyway. In this context it is interesting to note that at least one developing country Tourism Board, that of Sarawak in East Malaysia, now aims for tourism development that protects the Borneo rainforests, but deliberately avoids the term 'ecotourism' in its marketing effort (Johnson, 1998). Further, the essential requirement for the achievement of sustainable development is that all tourism be made as sustainable as possible, not that more tourists spend their time marvelling at nature. A respectable case can even be made that sustainability may sometimes be better promoted by the construction of *artificial* destinations (Roberts, 1998), rather than by facilitating visits to the real thing.

Among the more obvious environmental impacts of tourism are waste generation and pollution, damage to coral reefs, mangroves, dunes, historic buildings and monuments, erosion, deforestation, and so on. Less obvious effects include contributions to ozone depletion and global warming, the economic marginalization of traditional users of protected areas, destruction or trivialization of traditional lifestyles (Hong, 1985; Croall, 1995), and increases in crime and prostitution. Additionally, many of tourism's advantages come with strings attached. For example, a large proportion of hard currency spending in developing countries finds its way back to the developed world as payment for imported goods and services. Management jobs in tourism in developing countries are often held by foreign nationals, while those tourism employment opportunities which are created for locals are frequently seasonal and may alter the labour economics of the domestic agricultural sector (Lawrence, 1994). Finally, evaluations of sustainability are complicated by issues of scale. For example, what is judged to be acceptable and appropriate as a sustainable initiative nationally, may prove disastrous for a particular locality. Cost and revenue effects of changes in the environmental behaviour of tourism firms vary across different timescales, and energy savings or emission reductions per tourist may still mean increased energy use and pollution as total tourist volumes rise (Goodall, 1995).

Education, training and tourism

There is quite wide agreement that education and training are important to the achievement of sustainable tourism (Ham et al., 1991; Cater and Goodall, 1992; Johnson, 1998) and sustainable development (UNESCO–UNEP, 1996). Unfortunately, there is much less agreement about who should learn what, from whom, and how. This is perhaps unsurprising, given that tourism probably touches, at all levels, upon a wider range of social interests and economic sectors than any other industry (Cater, 1995) and that most academic disciplines have a bearing of some sort on the question of how to make tourism sustainable.

Ham et al. (1991) identify four key audiences for environmental *interpretation* in developing countries; subsistence-level locals, upper and middle-class nationals, influential nationals, and foreign tourists. They make the point that the environmental learning needs of each of these groups are different. Taking a broader view of tourism, a number of other target groups for sustainable tourism education and training suggest themselves, including employees and managers in the hospitality, travel and construction industries, government officials in host countries, potential tourists at their point of origin and, of course, children at school who may assume these or other roles in the future. Each of these groups is likely to have a different perspective on the environment and sustainability, and a different expectation of what education might offer them. Many are likely to believe that they have more to teach than to learn. However, different groups have different access to the power and resources to enable them to put their own analysis forward, and are likely to marshal definitions of terms, moral arguments and scientific evidence in ways which support their existing view of the problems and priorities. Environmental educators and trainers, therefore, must address a heterogeneous audience among which a variety of preconceived ideas relating to tourism and tourism development are likely to be held. Further, it seems clear that uncertainty and contestation, even about the meaning of 'sustainable tourism', are likely to persist for the foreseeable future. Equally clearly, environmental education and training which helps to create and support sustainable tourism is needed now.

Environmental education: determining an approach

A further problem is that, though a tentative consensus has been reached on the desirability of certain characteristics for environmental education – for example holism (Sterling, 1993), a life-long educational approach (UNESCO–UNEP, 1977), a consideration of values as central to any programme (Caduto, 1985), and a focus on the future (Hicks, 1996) – there is nevertheless substantial disagreement about what exactly the goals of environmental education should be (for example, Hungerford and Volk, 1990; Fien, 1993) and how they might relate to sustainable development (Huckle and Sterling, 1996). For example, the goal of 'environmentally-affirmative citizenship' (Hungerford and Volk, 1984) may seem useful as a basis for the environmental education of domestic tourists in rich countries and also, perhaps, of local tourism managers in developing countries. Socially-critical environmental education, which rejects this approach and emphasises instead egalitarianism, political decentralization, the reversal of industrialization and growth, and the abandonment of consumerism, may be appropriate elsewhere. Critically-theorized action research has lead to successful educational and political direct action in countries with a liberal tradition, for example to protect beaches (Greenall Gough and Robottom, 1993). Such action research may also be an effective method for extension work with displaced indigenous groups (Gough, 1995). However, neither approach seems wholly appropriate, by itself, to situations of the kind in which, say, a multinational hospitality concern obtains political approval in a developing country to set up a joint-venture with local entrepreneurs, and engages a foreign contractor to develop a major international resort using a labour mix of local people, nationals from other regions of the country and immigrants on short-term contracts. This is particularly so, given the observation of Jensen and Schnack (1997), that environmental education initiatives should be evaluated on the basis of their educational value to learners rather than their success in solving society's problems. In such a situation potential learners have different starting points and different expectations.

Cultural theory: developing a clumsy approach

There seems, therefore, to be a need for a theoretical device which permits environmental education processes to continue in the face of contradictory perceptions of their purpose. Such a device is 'cultural theory' (Thompson, Ellis and Wildavsky, 1990) which has been developed from the work of the cultural anthropologist, Mary Douglas (1982).

An approach from cultural theory starts from the observation that human knowledge, both of the natural environment and of human interactions with it, is imperfect and characterized by uncertainty and risk (James and Thompson, 1989; Thompson, 1990). In the face of this uncertainty and risk, social actors construct their interpretations of environmental reality. It is instructive to visualize such interpretations as lying within a range bounded by four archetypes: the fatalistic interpretation, the hierarchical interpretation, the individualistic interpretation and the egalitarian interpretation. These archetypes, in turn, represent possible combinations across two dimensions of social organization: equality/inequality and competition/no competition (Figure 4.1). Hence, the fatalist visualizes the social and natural worlds as competitive and unequal, but for the individualist they are competitive and equal; and while the expectations of the hierarchist are built upon assumptions of inequality and uncompetitiveness, those of the egalitarian assume, rather, uncompetitiveness with equality. Each archetype is further associated with a particular 'myth of nature'. For the fatalist, nature is capricious; for the individualist, it is benign. Hierarchists suppose nature to be benign within certain limits, but perverse if those limits are exceeded. Finally, egalitarians view nature as ephemeral; a delicate equilibrium which may be easily and irretrievably destroyed.

Which interpretation and myth of nature an individual is likely to favour depends on whether his or her 'social solidarity' (Thompson, 1997, p. 144) in the specific context under consideration, is individualist, hierarchist, egalitarian or fatalist. Solidarities, and therefore interpretations, may shift repeatedly over time and in response to changes of social context such as that from, say, workplace to family home.

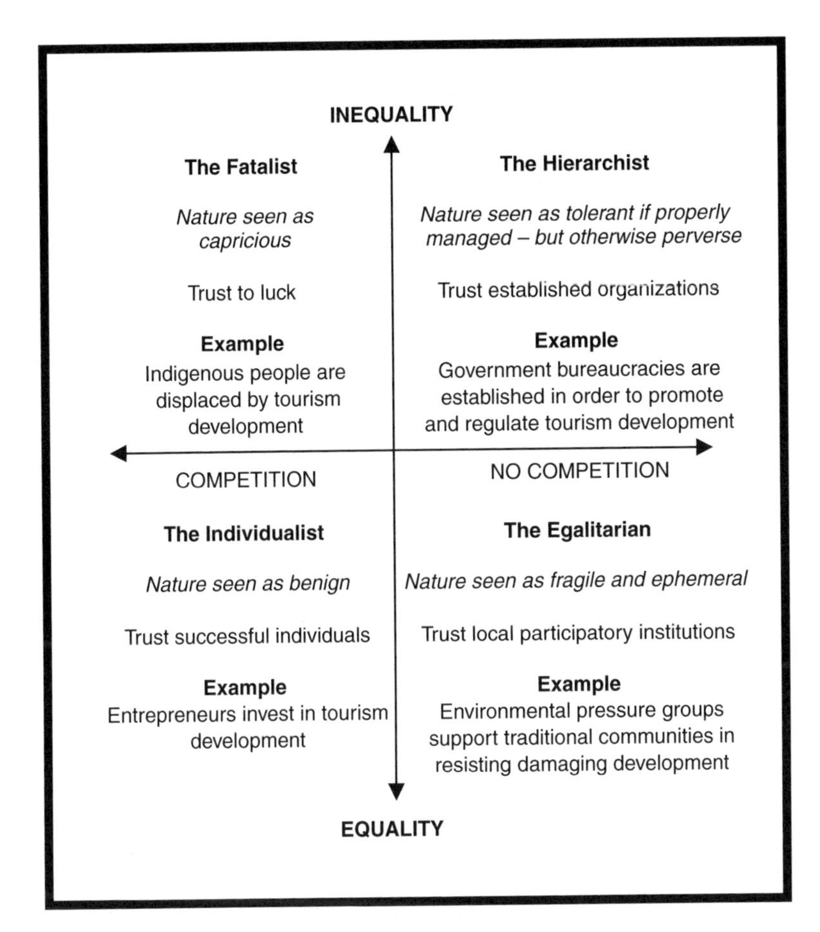

Figure 4.1 Four archetypal constructions of environmental reality
Source: Adapted from James and Thompson, 1989.

However, each of the four is seen as defining itself in contradistinction to the others and therefore of being incapable of sustaining itself without them, that is each view needs a perceived threat from the others if it is itself to make sense. Thompson writes:

> Humans, individualists know, are *self-seeking*; hierarchists know they are *malleable* (born in sin but redeemable by firm and nurturing institutions); egalitarians know that they are *caring and cooperative* until corrupted by coercive institutions (markets and hierarchies); fatalists know they are *fickle*. (Thompson, 1997, p. 145, original emphasis)

Hence disagreement, competition and even conflict between rival individuals and groups is not a social aberration but, on the contrary, an essential characteristic of society's uncertain relationship with its environment. Useful interventions are therefore most likely to originate with 'clumsy institutions' (Thompson, 1990), which, rather than being committed to one of the four perspectives, are prepared to entertain all of them. The reward such institutions may reap for tolerating a degree of apparent internal inconsistency is the discovery of synergies between opposed views.

From the points of view of ethnographic educational research, curriculum theory and environmental education, the idea of socially-constructed, context-specific knowledge is not at all new (Wolcott, 1988; Kemmis, Cole and Suggett, 1983; Fien, 1993), but cultural theory argues that the ways in which knowledge may be constructed are not infinitely variable. Rather, they derive from interplay between the four ideal types of social organization. Schwarz and Thompson write: An act is rational if it supports a person's way of organizing (Schwarz and Thompson, 1990, p. 61).

This analysis seems likely to be useful in designing and introducing an educational innovation under complex social circumstances (such as those attending sustainable tourism development) if it enables anticipation and prior classification of (at least) most of the range of eventual responses. A further attraction is the ability of a clumsy approach grounded in cultural theory to accommodate variations in spatial and temporal scale (Thompson, 1997), that is, to recognize that an individual's solidarities may be expected to vary not only with cultural origins, employment and education, for example, but also across his or her relations within, say, the household, extended family, village, place of religious worship, dialect group, locality, region of the country, country, and geopolitical region. Further, social solidarities may be expected to change as a person ages, and according to whether the short, long or medium term is being considered. A possible example of a particular dissonance of this type is between, on the one hand, the long-term hopes and expectations of students in developing countries who enrol for tourism-related training, and on the other the inevitably more short-term (and progressively shortening) perspectives of expatriate experts employed to deliver such training on fixed-term contracts. Thus, this clumsy approach seems particularly appropriate for use in, say, management education in a developing country context.

Adaptive concepts

An innovative method of clumsy intervention in the educational problem of sustainable tourism development has involved the use of 'adaptive concepts'. An adaptive concept is an enabling idea which holds the possibility of being able to facilitate discourse between disparate, or possibly hostile individuals and groups by appealing to holders of different interpretations of the environmental reality of a particular setting. It is an idea, a property or a value which has an established importance in environmental education theory and practice, and is also recognizably significant within the literatures of environmental management and economic development. Thus, an adaptive concept can 'resonate' between fields and groups, and across scales, in a way which allows environmental education to 'plug into' areas where normally its focus and concerns are not found. For example, against a socially and environmentally-complex and uncertain background of tourism development it might be found that: government bureaucracies favour a hierarchical, managerial and procedural approach to the achievement of sustainability; environmental pressure groups offer an egalitarian account which sees changes in human ethics as essential to avoid imminent environmental catastrophe; international lending institutions advocate free-market solutions based in an individualistic and competitive view of rationality; and, finally and in spite of all this concern for their well-being, local people slip gradually into fatalism. Hence, in *designing* a programme of (eco)tourism development, hierarchists will probably begin by establishing an official Tourism Board, egalitarians by exploring local environmental knowledge, and individualists by researching customer requirements. Meanwhile, local people are likely to have their own ideas about exactly what changes would lead to an improvement in their circumstances, and may be confused by, say: simultaneous exhortations both to move with the times and conserve tradition; the prospect of giving up a traditional lifestyle in exchange for a job telling foreigners what it was like; the gradual replacement of natural cycles by economic cycles as the basis of their pattern of life. An adaptive concept needs to be recognized as meaningful and significant by all these interests, and to be able to accord them all credibility through its use in teaching and learning situations.

Using the adaptive concept 'design'

Some evidence as to the potential of adaptive concepts in general in environmental education for sustainable tourism, and that of the adaptive concept 'design' in particular, was provided by a small, two-phase research project which sought to develop, use, and evaluate an environmental education approach appropriate to an internationally-available, pre-university management curriculum. The formal enviromental content of this programme specified by the examining board was nil. Materials were designed for use by experienced management teachers from the UK, Canada, New Zealand and Australia with a cohort of approximately 170 Malay and Chinese students aged from 16 to 21 years. The setting for the research was a college in Brunei, North Borneo, from which management students were very likely to proceed to further studies or employment in business and administration (Murshed, 1995). Tourism had not been the initial focus of the project, but had emerged strongly as a theme of interest to students during its first stage, which explored the use of another adaptive concept, 'quality' (Gough, Oulton and Scott, 1998). During the second phase tourism became central. Groups with a direct interest in the progress of the work at all or some of its stages were:

- management students
- management teachers
- non-management students and teachers
- parents
- college administrators
- government officials
- the business community.

The main focus of both teaching and research was a week-long activity in which students, working in groups, were required to design a marketing mix suitable for the promotion of tourism in Brunei. The concept of a marketing mix is fundamental in management education at this level. Tourism had been targeted locally as a potential growth industry which was central to national development planning. Students were encouraged to consider local tourism development issues both in general terms and with a particular focus on the country's coastal regions, which are ecologically rich but under pressure for development from a variety of economic directions.

The purposes of this tourism marketing-mix design intervention fell into two broad categories, one concerned with teaching and learning, the other with research. In terms of teaching and learning the purposes were:

- to enhance students' management training
- to show evidence of achievement of environmental education goals from the perspectives of more than one educational research paradigm
- to show evidence of achievement of professional development of teachers in environmental education from the perspectives of more than one educational research paradigm.

In terms of research the purposes were:

- to evaluate the use of the adaptive concept 'design' as a means of introducing environmental education into this management programme
- to continue evaluation (which had begun with the use of the adaptive concept 'quality') of the usefulness of adaptive concepts in general as a means of focusing both curriculum development and research in environmental education
- to achieve interpretivist understanding and illumination of the context of the intervention.

The following methods were employed during the teaching activities which provided research data. Students were provided with a booklet containing information drawn from official publications, the local press, the management literature on tourism, the environmental literature on tourism, documents supplied by local firms whose work had environmental impacts, and extracts from the written output of students during the first phase of the research using the adaptive concept 'quality'. Students were asked to complete two tasks bearing on the future design of Brunei's tourism product. The first was to prepare a report containing recommendations for the usage of the local coastline over the next six years which balanced government economic, developmental, environmental and cultural objectives with a variety of interests including those of tourism firms, local villagers, local fishermen, shipping firms and wildlife. The second was, in the light of that report, to identify target markets for local tourism development, make recommendations regarding the types of tourism and tourist to be encouraged, outline a preferred usage pattern for local resources, and discuss potential costs and benefits of all kinds. Subsequently, groups of volunteer students continued and refined these studies with the researcher.

It was considered important to meet or exceed the expectations of business education held by parties to the research while simultaneously maintaining, and in fact propagating, an environmental education focus. To this end educational administrators were regularly consulted and informed of progress. Arrangements were made for students to present work to the Head of the Tourism Unit at the Ministry of Industry and Primary Resources, and to further discuss it with him during a seminar held at the nearby ASEAN–EU Management Centre. All teachers involved completed a detailed written evaluation of the intervention. The selection and use of illustrative examples in a sample of students' subsequent syllabus-specific work were examined to investigate the extent to which students were showing evidence of enhanced understanding of the relevance of environmental issues to business management, and at the same time meeting external requirements.

To achieve the research purposes of the intervention, data about students' and teachers' responses to the teaching activities were collected in the forms of:

- students' written proposals
- this researcher's field notes
- audio-taped conversations between students and this researcher
- teachers' written evaluations.

The intervention was evaluated against environmental education and professional development guidelines from both positivist (Hungerford, Peyton and Wilke, 1980; Hungerford and Volk, 1990) and critical (Fien, 1993; Robottom, 1987) educational orientations. This was felt to be consistent both with the 'clumsy' commitment of the research to respect conflicting perspectives under conditions of uncertainty and, more fundamentally, with a cultural-theory-based understanding of the nature of the disagreements between different educational methodologies. Positivist research in environmental education, with its emphasis upon the role of the citizen, assumes a target audience of networking individuals. Critical theory is based upon egalitarian assumptions about the social world it addresses. Confusingly, however, academics from both paradigms typically conduct their debates within the hierarchical frameworks of higher education and research.

Emphasis was placed on the written data supplied by students and teachers. In the case of the students this was for two reasons. First, the teaching activities involved a focus on possible future outcomes, and the literature of futures studies in environmental education suggests that, 'Pupils' interests become more alive when they *write* about their hopes and fears for the local area' (Hicks and Holden, 1995, p. 188, original emphasis). Second, since all students were responding in their second (at least) language, it was felt appropriate to permit them the opportunity to consider, and if necessary correct, what they wanted to say. In the case of teachers, written data were required for research purposes because, once more, a considered response was being sought, and because to provide a written evaluation of a curriculum innovation would be a normal and expected thing for a teacher to do in this setting.

Written data from students were analysed and checked for internal validity using techniques drawn from network analysis (Bliss, Monk and Ogborn, 1983), problem-based methodology (Robinson, 1993) and dilemma analysis (Winter, 1982). A diagrammatic representation of the ideas which each student thought important for tourism development, and the way in which these were interconnected, was drawn up following simple rules to relate linkages made by students to the configuration of the finished diagram. For example, if a student expressed the view that damage to mangroves would lead to a decline in traditional small scale fishing activities, then a cause and effect relationship was recorded by linking 'damage to mangroves' to 'decline of traditional fishing' by means of a descending vertical line. On the other hand, if both damage to mangroves and the decline of fishing were considered by the student to be consequences of the construction of hotels this was recorded by linking 'damage to mangroves' to 'decline of traditional fishing' horizontally on the page, with a further vertical link upwards to 'hotel construction'. The term 'loose networks' was coined for the resulting research artefacts, to emphasize that the ideas they contained, and the inter-relationships between them, originated with the students, not the researcher (Figure 4.2). However, the researcher's concerns became significant at the next stage of data analysis, the identification of 'classes of interest' (Robinson, 1993, pp. 125–6) from the loose networks. Students' ideas became interesting if they (or the links made between them) seemed likely to have a bearing on the achievement of the purposes of the research. There was, then, 'interplay between the theorising of outsiders and insiders in the development of these classes' (Robinson, 1993, p. 125). Examples of classes of interest included: the need to preserve traditional customs; concern for traditional activities such as fishing and the manufacture of musical instruments; hopes for future prosperity; national and international issues of social justice; the merits of the expansion of the built environment at the expense of jungle and mangroves; and the proper role of public policy.

Wherever there appeared to be tension within a class of interest this was formulated as a 'dilemma'. A total of 65 such dilemmas were expressed as pairs of statements. These were then assembled into a 'perspective document' (Winter, 1982) in the form of a questionnaire, which provided both a check on the validity of the loose networks and a potential source of further adaptive concepts (Figure 4.3). Where a large proportion of students (arbitrarily chosen as over 65 per cent of those responding) expressed explicit agreement with both statements in a pair, the focus of the statements was deemed to be strengthened as a class of interest, and became a candidate for consideration as an adaptive concept. Examples of issues addressed in this way included: the tendency of tourism to destroy its own attractions; traditional versus modern technologies;

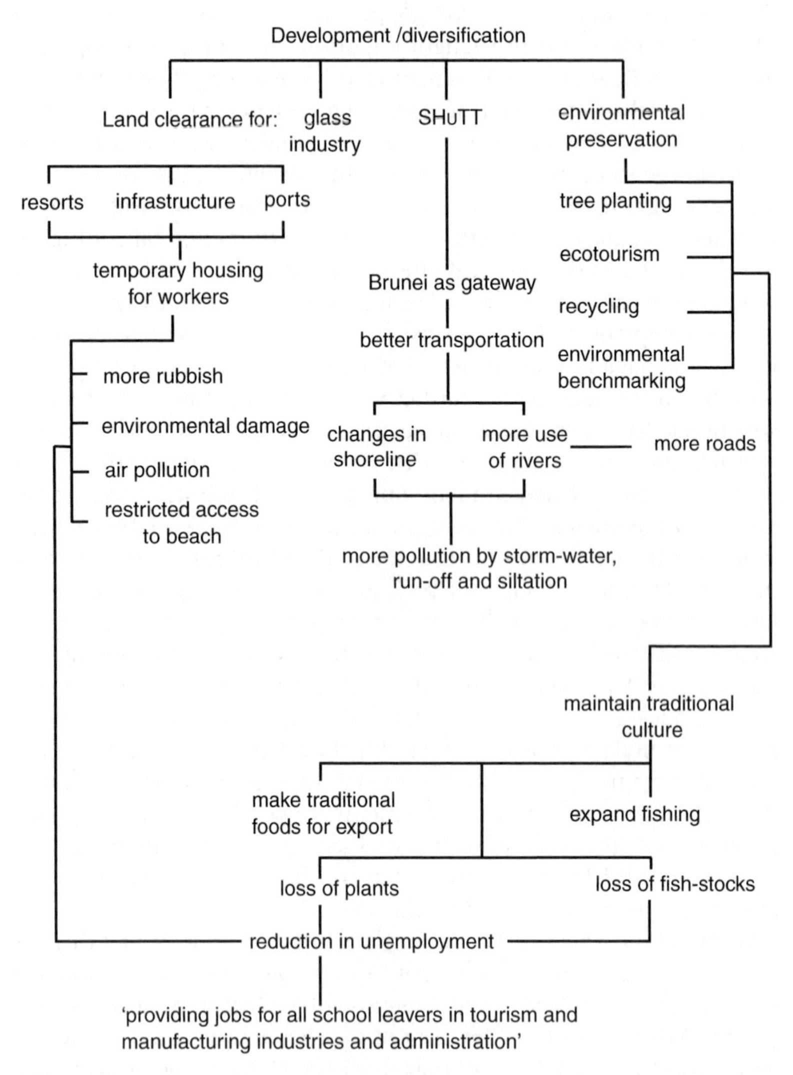

Figure 4.2 Example of a 'loose network'

Note: 'ShuTT' is an acronym used to describe a development plan under the terms of which Brunei would become the 'Service Hub for Trade and Tourism' within a designated south-east Asian regional growth area (BIMP-EAGA), by the year 2020.

100	Part of Brunei's tourism appeal is that it is the 'Abode of Peace'. Brunei is peaceful because it receives few tourists.
101	Brunei must change to suit tourists. Brunei is attractive to tourists as it is.
102	Tourism is a way of diversifying from oil and gas. Brunei's oil and gas wealth is part of its appeal to tourists.
103	Natural habitats must be preserved. Roads and railways must be built through natural habitats so that tourists can see them.
104	Tourists must be able to enjoy themselves as they wish. Tourists must respect Bruneian values and traditions.
105	Islands in Sg. Brunei are valuable because they provide an undisturbed environment. Islands in Sg. Brunei are valuable because they can be exploited and managed.
106	Muara and Serasa should provide beach attractions to tourists. Muara and Serasa should provide industrial and commercial facilities.
107	Tourists are attracted by traditional activities like fishing, craft work and cooking Jobs in tourism will replace traditional ways of earning a living.
108	Tourists will not visit areas where the natural environment has been damaged. Tourists are a major cause of damage to the natural environment.
109	Tourism revenues will replace oil and gas revenues in the long run. Tourist expenditures are often on imported goods and services.
110	Tourist numbers must be kept low to minimize damage to coastal and jungle ecosystems. Tourist numbers must be maximized to earn the greatest possible revenue.

Figure 4.3 Sample of pairs of statements included in the 'perspective document'

Note: 'Serasa' and 'Muara' are, respectively, a location near the mouth of the Brunei river which is home to a fishing community and an industrial estate, and the port at the mouth of the same river.

the varied factors which motivate tourists, tour-operators, tourism employees and local people; the meanings of the term 'resources'; the difficulties of valuing environmental assets; the proper place of traditional artefacts, dance and song; the finance of tourism; the allocation of benefits from tourism; issues of scale; and coastal and rainforest ecology as it affects and is affected by tourism development. Which classes of interest were subsequently to be adopted as adaptive concepts would depend upon their significance for other groups in the research context, viz: students, management teachers, college administrators and tourism development officials.

These same groups confirmed that the tourism design work discussed here was an appropriate *management* education innovation. The teachers were unanimous in wishing to continue the work with future cohorts, and discussions took place about the possibility of a further programme using other potential adaptive concepts which were identified during the research. Subsequently, evidence emerged that students were also raising these issues appropriately in their examination-related work in Management, and a number of individual students seeking admission to Higher Education institutions in the UK in commercial subjects drew specific and unprompted attention to their tourism marketing design work as part of their UCAS personal statements. Finally, the project had a practical spin off. It was instrumental in the initiation of work in another department which culminated in the establishment of a college recycling system.

Concluding comment

This small research effort claims educational value in its setting, no more. In particular, the institutional resistance or incomprehension often encountered by environmental education initiatives (Posch, 1991) was avoided, a degree of engagement of environmental education ideas was achieved within settings to which they might well be thought alien (Gough, 1997) and there appear to have been gains in terms of teaching and learning. Careful generalization may be possible from qualitative, case study research (Adelman, Jenkins and Kemmis, 1980; Robinson, 1993) but further work with adaptive concepts in different contexts is necessary before any such claims could be made for them.

Acknowledgement

This chapter was previously published in 1999 in the *Canadian Journal of Environmental Education*, 4, 193–212.

References

Adelman, C., Jenkins, D. & Kemmis, S. (1980) 'Rethinking case study: notes from the second Cambridge conference'. In H. Simons (ed.), *Towards a Science of the Singular*, pp. 47–61. Norwich: University of East Anglia, Centre for Applied Research in Education.

Alderman, C.L. (1994) 'The economics and the role of privately-owned lands used for nature tourism, education and conservation'. In M. Munasinghe &

J. McNeely (eds), *Protected Area Economics and Policy: Linking Conservation and Sustainable Development*, pp. 273–317. Washington, DC: World Bank/IUCN.

Bliss, J., Monk, M. & Ogborn, J. (1983) *Qualitative Data Analysis for Educational Research*. London: Croom Helm.

Caduto, M. (1985) *A Guide on Environmental Education Values Education*. Paris: UNESCO–UNEP.

Cater, E. (1995) Environmental contradictions in sustainable tourism. *The Geographical Journal*, 161(1), 21–8.

Cater, E. & Goodall, B. (1992) 'Must tourism destroy its resource base?' In A.M. Mannion & S.R. Bowlby (eds), *Environmental Issues in the 1990s*, pp. 309–24. Chichester: John Wiley.

Charter, M. (1992) 'Emerging concepts in a greener world'. In M. Charter (ed.), *Greener Marketing: A Responsible Approach to Business*, pp. 55–92. Sheffield: Greenleaf.

Croall, J. (1995) *Preserve or Destroy: Tourism and the Environment*. London: Gulbenkian Foundation.

Douglas, M. (ed.) (1982) *Essays in the Sociology of Perception*. London: Routledge & Kegan Paul.

Fien, J. (1993) *Education for the Environment: Critical Curriculum Theorising and Environmental Education*. Geelong: Deakin University Press.

Filion, F.L., Foley, J.P. & Jacquemot, A.J. (1994) 'The economics of global ecotourism'. In M. Munasinghe & J. McNeely (eds), *Protected Area Economics and Policy: Linking Conservation and Sustainable Development*, pp. 235–52. Washington, DC: World Bank/IUCN.

Goodall, B. (1995) 'Environmental auditing: a tool for assessing the environmental performance of tourism firms'. *The Geographical Journal*, 161(1), 29–37.

Gough, S. (1995) 'Environmental education in a region of rapid economic development: the case of Sarawak'. *Environmental Education Research*, 1(3), 327–37.

Gough, S. (1997) 'Adding value: an environmental education approach for business and management training'. *Environmental Education Research*, 3(1), 5–16.

Gough, S., Oulton, C.R. & Scott, W.A.H. (1998) 'Environmental education, management education and sustainability: exploring the use of adaptive concepts'. *International Journal of Environmental Education and Information*, 17(4).

Greenall Gough, A. & Robottom, I. (1993) 'Towards a socially critical environmental education: water quality studies in a coastal school'. *Journal of Curriculum Studies*, 25(4), 301–16.

Ham, S.H., Sutherland, D.S. & Meganck, R.A. (1991) *Taking Environmental Interpretation to Protected Areas in Developing Countries: Problems in Exporting a US Model*. Mimeograph, College of Forestry, Wildlife and Range Sciences. Idaho: University of Idaho.

Hicks, D. (1996) 'Envisioning the future: the challenge for environmental educators'. *Environmental Education Research*, 2(1), 101–8.

Hicks, D. & Holden, C. (1995) 'Exploring the future: a missing dimension in environmental education'. *Environmental Education Research*, 1(2), 185–94.

Hong, E. (1985) *See the Third World While It Lasts: The Social and Environmental Impact of Tourism with Particular Reference to Malaysia*. Penang: Consumers Association of Penang.

Huckle, J. & Sterling, S. (1996) *Education for Sustainability*. London: Earthscan.

Hungerford, H., Peyton, R. & Wilke, R. (1980) 'Goals for curriculum development in environmental education'. *Journal of Environmental Education*, 11(3), 42–7.

Hungerford, H. & Volk, T. (1984) 'The challenges of K-12 environmental education'. In A.B. Sacks (ed.), *Monographs in Environmental Education and Environmental Studies*, vol. 1, pp. 5–30. Troy, Ohio: NAAEE.

Hungerford, H. & volk, T. (1990) 'Changing learner behaviour through environmental education'. *Journal of Environmental Education*, 2(3), 8–17.

James, P. & Thompson, M. (1989) 'The plural rationality approach'. In J. Brown (ed.), *Environmental Threats: Perception, Analysis and Management*, pp. 87–94. London: Belhaven Press.

Jensen, B.B. & Schnack, K. (1997) 'The action competence approach in environmental education'. *Environmental Education Research*, 3(2), 163–78.

Johnson, R. (1998) 'Putting the eco into tourism'. *Asia Magazine*, 36(13), 8–12.

Kemmis, S., Cole, P. & Suggett, D. (1983) *Orientations to Curriculum and Transition: Towards the Socially-Critical School*. Melbourne: Victorian Institute for Secondary Education.

King, V.T. (1993) 'Tourism and culture in Malaysia'. In M. Hitchcock, V.T. King & M.J.G. Parnell (eds), *Tourism in South East Asia*, pp. 99–116. London and New York: Routledge.

Lawrence, K. (1994) 'Sustainable tourism development'. In M. Munasinghe & J. McNeely (eds), *Protected Area Economics and Policy: Linking Conservation and Sustainable Development*, pp. 263–9. Washington, DC: World Bank/IUCN.

Mowforth, M. (1993) 'In search of an ecotourist'. *Tourism in Focus*, 9, 2–3.

Murshed, H. (1995) 'Management education and training in Brunei Darussalam'. *Country Reports: Management Education and Training*, pp. 1–20. Bandar Seri Begawan: Universiti Brunei Darussalam and Association of South East Asian Institutions of Higher Learning.

Pleumarom, A. (1994) 'The political economy of tourism'. *The Ecologist*, 24(4), 142–8.

Posch, P. (1991) 'Environment and schools initiative: background and basic premises of the project'. *Environment, Schools and Active Learning*, pp. 13–18. Paris: Centre for Educational Research and Innovation, OECD.

Roberts, M. (1998) 'Dream factories: a survey of travel and tourism'. *The Economist*, 10 January, supplement.

Robinson, V.M. (1993) *Problem-Based Methodology*. Oxford: Pergamon Press.

Robottom, I. (1987) 'Towards inquiry-based professional development in environmental education'. In I. Robottom (ed.), *Environmental Education, Practice and Possibility*, pp. 83–120. Geelong: Deakin University Press.

Schwarz, M. & Thompson, M. (1990) *Divided We Stand: Redefining Politics, Technology and Social Choice*. Philadelphia: University of Pennsylvania Press.

Sterling, S. (1993) 'Environmental education and sustainability: a view from holistic ethics'. In J. Fien (ed.), *Environmental Education: A Pathway to Sustainability*, pp. 69–98. Geelong: Deakin University Press.

Thompson, M. (1990) *Policy Making in the Face of Uncertainty*. London: Musgrave Institute and Wye College; Geneva: International Academy of the Environment.

Thompson, M. (1997) 'Security and solidarity: an anti-reductionist framework for thinking about the relationship between us and the rest of nature'. *The Geographical Journal*, 163(2), 141–9.

Thompson, M., Ellis, R. & Wildavsky, A. (1990). *Cultural Theory*. Boulder, Colorado and Oxford: Westview.

Tickell, C. (1993) Quoted in: *The Geographical Journal*, 159(1), 114.

UNESCO–UNEP (1977) 'The Tbilisi Declaration'. *Connect*, 3(1), 1–8.

UNESCO–UNEP (1996) *Connect*, XXI(2), 1–3.

Winter, R. (1982) 'Dilemma analysis: a contribution to methodology for action research'. *Cambridge Journal of Education*, 12(3), 161–174.

Wolcott, H.F. (1988) 'Ethnographic research in education'. In R.M. Jaeger (ed.), *Complementary Methods for Research in Education*, pp. 187–206. Washington, DC: American Educational Research Association.

Part III

Sustainability and Resources

5
Cities and Sustainability

Mark R.C. Doughty and Geoffrey P. Hammond

Summary

In the aftermath of the Brundtland Report and the 1992 Rio Earth Summit the concept of sustainability has become a key idea in national and international discussions. Sustainable development is certainly a desirable and, more debatably, an attainable objective in global terms. However, it is less obviously applicable on a smaller scale, where it is sometimes used synonymously with self-sufficiency. It is argued that the notion of 'sustainable cities', popularized in contemporary literature, is simply based on a misconceived idea of the full implications of sustainability, as well as the way that cities have developed historically. The technique of environmental footprint analysis is used to examine the sustainability of cities by placing them in their broader geographic context. The eighteenth-century (Georgian) city of Bath is employed as a case study following Doughty and Hammond (1997 and 2000). Its per capita 'environmental footprint' is contrasted with that of the surrounding bioregion, as well as with the United Kingdom, Europe and the World. The footprint of Bath is found to be greater than that of these wider geographic regions, and is nearly twenty times larger than its corresponding land area. This lends support to the authors' critique of the idea of sustainable cities; cities only survive because they are linked by human, material and communications networks to their hinterlands or bioregions. Nonetheless, the urban design of compact cities can obviously contribute to a more sustainable way of life, particularly in industrialized societies. This can be done by encouraging the development of integrated, mixed-use, urban communities as has been advocated by a diverse range of architectural critics. But they must be seen as part of a wider spatial matrix.

Introduction

'Sustainable cities'

The concept of sustainability has become a key idea in national and international discussions following publication of the Brundtland Report (WCED, 1987) and the 1992 Rio 'Earth Summit'. It has been given further prominence in the run-up to the 2002 Rio+10 review conference being held in Johannesburg. Sustainable development is certainly a desirable and, more debatably, an attainable objective in global terms. However, it is less obviously applicable on a smaller scale, where it is sometimes used synonymously with self-sufficiency. The Latin root of the word 'civilization' is *cives* (Hall, 1998) – citizen – and so cities are clearly at the heart of human development. However, it is currently fashionable in literature to find material on 'sustainable cities' (see, for example, Giradet, 1992 and 1999; Haughton and Hunter, 1994; Rogers, 1997, and the UK Urban Task Force, 1999). Indeed, part of the UK Research Councils' Clean Technology Managed Programme was devoted to this topic (see Ekins and Cooper, 1993). Nevertheless, it is argued here that this notion is quite spurious in systems analysis terms (see also Day and Hammond, 1996). It looks back to the 1970s' idea of self-sufficiency, when it became popular to strive for 'autarkic' buildings or settlements. Clusters of buildings and an integrated human-scale transport infrastructure can enhance energy conservation and reduce environmental impact. But even 'compact cities' are not in themselves sustainable. They survive only because they are inextricably linked by human, material and communications networks to their hinterlands or bioregions. This outlying support structure extends from the regional to national and even global scale.

The problem considered

In order to examine the evolution of contemporary views on the role of cities in the wider economy, their historical development in ancient and medieval times is reviewed. This highlights the benefits of urban communities, as well as their dependence on the surrounding bioregions. Utopian visions of urban habitats stretching from the level of individual buildings to that of whole settlements are also examined. These visions developed over recent decades have acted as precursors for the notion of sustainable cities as popularized in the modern architectural and urban studies literature.

The technique of environmental footprint analysis is used to illustrate the role of cities in aiding sustainability by placing them in their broader context. Thus, the resource and environmental impact of the city will be examined in wider geographic spheres. Bath provides the central focus of the study, together with its ever-expanding links with the former County of Avon, the South West, the United Kingdom, Europe, and the World. This will yield a snapshot of sustainability issues, based on the footprint analysis of Doughty and Hammond (1997 and 2000). The uncertainties and deficiencies of using environmental footprints (and related parameters) as sustainability indicators are highlighted, including problems of urban boundary definitions, data gathering, and the basis for weighing the various consumption and associated impacts.

Sustainable development versus sustainability

Over a period of some 15–20 years, the international community has been grappling with the task of defining the concept of 'sustainable development'. It came to particular prominence as a result of the so-called Brundtland Report published in 1987 under the title 'Our Common Future'; the outcome of four years of study and debate by the World Commission on Environment and Development (WCED, 1987) led by the former Prime Minister of Norway, Gro Harlem Brundtland. This Commission argued that the time had come to couple economy and ecology, so that the wider community would take responsibility for both the causes and the consequences of environmental damage. It defined sustainable development as meeting 'the needs of the present without compromising the ability of future generations to meet their own needs'. It therefore involves a strong element of intergenerational ethics. Many writers and researchers have acknowledged that the concept of 'sustainable development' is not one that can readily be grasped by the wider public (see, for example, Hammond, 1998, 2000b and 2001). However, no satisfactory alternative has been found. The UK Government has sometimes adopted the layman's term 'quality of life' as a shorthand expression when referring to sustainable development issues applicable to both Britain and the wider world (DETR, 1999). However, the former UK Round Table on Sustainable Development (1999), one of the precursors of the current Sustainable Development Commission, amongst others, argued that the quality of life is only part of what is meant by sustainable development. They contended that it would be unfortunate if the two expressions became synonymous in the 'public mind'.

The notion of sustainable development is not without its critics. Meredith Thring (private communication, 1999) regards the term as an oxymoron; arguing that development *per se* cannot be sustainable.

He would prefer humanity to strive for a creative and stable world with the aid of 'equilibrium engineering' (Thring, 1990). Similar views can be found in developing countries, where their debt burden and inequalities in global income distribution are seen as serious obstacles to sustainable development (Amin, 1997). On a more fundamental level, Jonathan Porritt (2000) has stressed that such development is only a process or journey towards a destination which is 'sustainability'. This process cannot easily be defined from a scientific perspective, although he argues that the attainment of sustainability can be measured against a set of four 'system conditions'. Porritt draws these from 'The Natural Step' (TNS); an initiative by the Swedish cancer specialist, Karl-Henrick Robèrt:

- Condition 1: Finite materials (including fossil fuels) should not be extracted at a faster rate than they can be redeposited in the Earth's crust.
- Condition 2: Artificial materials (including plastics) should not be produced at a faster rate than they can be broken down by natural processes.
- Condition 3: The biodiversity of ecosystems should be maintained, whilst renewable resources should only be consumed at a slower rate than they can be naturally replenished.
- Condition 4: Basic human needs must be met in an equitable and efficient manner.

These sustainability conditions put severe constraints on economic development, and they may therefore be viewed as being utopian (Hammond, 2001). They certainly imply that the ultimate goal of sustainability is rather a long way off when compared with the present conditions on the planet.

Historical development of cities

Ancient cities

Modern archaeological evidence suggests that the human species (*Homo Sapiens*) originally evolved from its close relatives around 150 000 years ago in East Africa. Skeletal remains of the genus *Homo* have been found along the Rift Valley stretching from Olduvai Gorge, in what is now Tanzania, to the shores of Lake Turkana in northern Kenya. These early hunter–gatherers ultimately spread out from their African 'birthplace' across the globe, reaching as far as the southern Americas by around 9000 BC (Parker, 1986). Human communication skills and the benefits of co-operative living meant that they eventually began to form settlements. Most cities develop over a long period, and their form depends on many factors; geographic, economic and cultural (Day and Hammond, 1996). Many of the early civilizations of the 'Old World' settled along the banks of rivers; along the Nile, around the Tigris and Euphrates, and in the Indus Valley. These three great civilizations are notably to be found in a narrow geographic band around 30 degrees latitude, where a largely temperate climate persists. A separate civilization, in what appears to have been a completely independent development, arose in the area of the Yellow River in China.

A classic example of an early 'New World' settlement is the Inca city of Machu Picchu in the High Andes of Peru (illustrated in Figure 5.1). This spectacular collection of structures was built around 1500 AD, near the time that Christopher Columbus first set foot on the West Indies. It had been abandoned for centuries when rediscovered by a young Western archaeologist, Hiram Bingham of Yale University, in 1911 (Bronowski, 1973). Machu Picchu exhibited a number of features that sustained it during the period that it flourished. It had a hinterland of rich agricultural land (with cultivated terraces and a system of irrigation based on canals and aqueducts), a strong central authority and a communications network with other cities. Such urban centres facilitate a community that is much richer and culturally diverse than are villages. They formed the foundation for the development of specialized skills and trades with, for example, goldsmiths, coppersmiths, weavers, and potters. Thus, Machu Picchu illustrates some of the basic requirements for sustainability: irrigation via canals and aqueducts to supply drinking water and facilitate the production of food (potatoes and maize were grown on the fertile terraces of its hinterland), a transport network based on crude roadways provided a link with other settlements, and a form of rudimentary communication was developed for message passing. Inca roads were long established, utilizing bridges with beams rather than arches, although they lacked any technological knowledge of the fixed-axle wheel. The Incas had no written form of language, and so message passing was achieved by way of a system of knotted strings called *quipus*. However, this was limited to recording numbers alone; something that does not seem to have restricted the Incas unduly, according to incomparable scientist and chronicler Jacob Bronowski (1973).

The tight social structure of the Inca civilization was unfortunately very fragile. It was quickly destroyed when its sun-god or king (the central authority of the Inca himself) was captured by invading Spanish conquistadors, who entered Peru in 1532. They were led by an illiterate

Figure 5.1 Ruins of the historic Inca city of Machu Picchu, High Andes, Peru
Source: Adapted from Bronowski (1973).

adventurer (Francisco Pizarro), and consisted of only some 106 foot-soldiers and 62 terrifying horses. All that now remains of Machu Picchu (Figure 5.1) are the basic masonry structures, now long overgrown, left empty in the Urubamba Valley of the eastern Andean mountains.

The medieval city

European cities of the 'Middle Ages' had both common and distinctive features. Most had a wall, a citadel (an inner fortification in case the outer one was breached), and a cathedral. In 1914 a German architect, Karl Gruber, prepared a set of drawings for his book *Ein Deuscher Stadt* that depicted a hypothetical twelfth- to eighteenth-century city in order to illustrate the similarities in urban development across the continent (Barnett, 1987). The 1180 version of Gruber's model city had a moat and a wall, winding streets of gabled houses, a citadel, and a cathedral facing the town hall across the market square. It was situated on the bend of a river, with a monastery on the opposite bank. By 1580 the fortifications of this hypothetical city are shown as being reinforced against primitive cannon on the battlements, following the introduction of gunpowder. Here the cathedral and castle had been reconstructed, but otherwise it is essentially the same city after four centuries. The last in the series of Gruber's drawings illustrates the model city in 1750, and rather more significant alterations. This suggests the need for greatly strengthened fortifications in response to more sophisticated cannons, and the reorganization of the city's urban profile. It is depicted as exhibiting a conscious urban plan, with squares and courtyards, yielding a symmetrical silhouette typical of the Baroque architectural style in contrast to the 'spiky' medieval one (Barnett, 1987).

British cities, including Bath and London, displayed several differences from their European counterparts in this period. They spread out beyond their early fortifications, relying for defence on Britain's naval power (England had not been successfully invaded by a foreign rival since William the Conqueror in 1066), and tended not to have a central market square. Instead they adopted a main street (Barnett, 1987), which performed much the same function as a market square in the UK context: the Milsom Street/Union Street/Stall Street 'ribbon' in Bath and Cheapside in London. Such thoroughfares are normally intersected by other streets to provide a network of shops and markets.

The origins of the heritage city of Bath in South West England (now within the unitary local authority of Bath & North East Somerset: B&NES), which is examined as a sustainability case study in 'Environmental footprint analysis' below, lay in its origins as a Roman settlement. Hot spring water that erupts from the ground was utilized by the Romans, both for bathing and for the central heating of their dwellings. They built the first hot baths in the first century AD, around the religious centre of *Aquae Sulis* (Woodward, 1992). However, the baths were largely disused after the withdrawal of the Romans in the fifth century, until

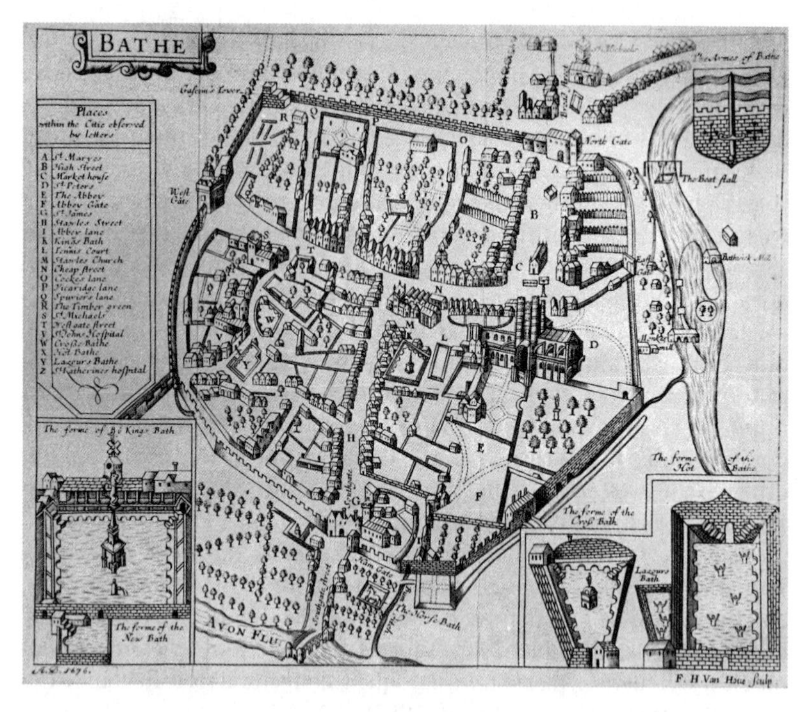

Figure 5.2 The medieval city of Bath c. 1676 AD

Source: Bath Historic Buildings Record; courtesy of Professor A.K. Day, Department of Architecture and Civil Engineering, University of Bath.

their supposed medicinal properties became more widely recognized in the sixteenth century as a cure for infertility and diseases such as leprosy and smallpox. Inevitably this led to the growth of the city in medieval times, as depicted in the contemporary (1676 AD) map shown in Figure 5.2. This mirrors a rather more famous map by the Bristol mathematician, Joseph Gilmore, in 1694 (see Woodward, 1992). In both, a mixture of baths, churches and houses, many constructed by local entrepreneurs to meet the needs of visitors, are clearly seen as being encapsulated by traditional battlement walls. The Gilmore map gives a slightly better feel for the surrounding farmland from which the agricultural requirements of the city were met. This configuration matches closely Gruber's hypothetical European model (c. 1180–1580), except for the absence of an inner citadel and a market square. Medieval Bath was constructed on a bend in the River Avon, with battlement walls and an abbey.

The city in the modern age

In contemporary society, cities house some 50 per cent of humanity across the globe (Rogers, 1997), out of a total population of a little over 6 billion in 2000. This represents a very rapid, virtually 'exponential', growth during the twentieth century, which opened with only 15 per cent of humans living in urban areas and a total population of about 1.65 billion. There are now 35 cities across the world with over 5 million people, and literally hundreds having more than 1 million (Girardet, 1999). This contrasts with 1800 AD when only London and Peking (renamed Beijing) had urban populations of 1 million. All cities provide access to community amenities and cultural events, but bring with them a range of social and environmental problems. The disadvantaged and minority groups tend to be concentrated in deprived inner city areas. Modern transportation systems, dependent as they are (to a great extent) on internal combustion engines, result in pollutant emissions and poor air quality, as well as the inevitable traffic congestion that bedevils major cities.

Modern cities exhibit a greater diversity, in many ways, than did their ancient and medieval counterparts (Barnett, 1987 and Hall, 1998). Some urban designers and planners have tried to characterize the different traditions. Jonathan Barnett (1987) divides the evolution of the city into four separate traditions: the monumental city (within which he includes both Bath and London), the garden city or suburb, the modern city (in architectural terms), and futuristic mega-structures. Sir Peter Hall, the distinguished (but idiosyncratic) geographer and urban planner, prefers to describe the rich and varied patchwork that constitutes the variety of cities that have developed on all continents throughout human history. He argues in his monumental work entitled *Cities in Civilization* (1998) that many have been primary sites for creativity and innovation. A favourite city of Hall's is Los Angeles; home of the 'movie' industry, but the first conurbation without a centre and one ruled by the car. Others might see this as more of a technological nightmare than an exemplar of some future idyll.

Utopian visions of urban habitat

Modern cities require vast amounts of resources, both for their urban inhabitants and for the economic activities concentrated there. They remain dependent on an ever-expanding hinterland to supply these resources every bit as much as did Machu Picchu or the medieval cities of Europe. Consequently, they cannot be viewed as sustainable in the limited sense of being self-sufficient; reliant on their own carrying capacity as a

Figure 5.3 A model of the 'autarkic' house

Source: The Autonomous Housing Project, Department of Architecture, University of Cambridge; design by Alexander Pike (c. 1975).

resource base. They could not, by themselves, meet the sort of 'sustainability' system conditions postulated by The Natural Step (Porrritt, 2000). Indeed, it could be argued that the first three conditions (see p. 84 above) are inevitably broken as a precondition of urban living.

Those who advocate greater self-sufficiency for cities, fostering (for instance) the development of city farms to provide food, are harking back to ideas for so-called 'autonomous', self-sufficient, buildings and settlements in the 1970s (Day and Hammond, 1996; Harper, 1976). These became fashionable in the aftermath of the oil crisis of 1972–3, when the need for greatly improved energy efficiency was widely acknowledged. An example of this movement is the 'autarkic' house

Figure 5.4 Escape from the cities; a new rural utopia

Source: *Undercurrents* 20 (1977)/Clifford Harper.

Figure 5.5 Renovating the cities; a response to the new rural utopia

Source: *Undercurrents* 21 (1977)/John Gilbert.

developed by Alexander Pike (c. 1975) in Cambridge (see Figure 5.3). This provided the design for a three-bedroom house with a roof-mounted aero-generator, and half the volume dedicated to a tall greenhouse. It was never actually constructed, although many admirers flew into Cambridge wanting to see the finished dwelling after having seen the pictorial representation (shown in Figure 5.3). Ideas of this sort led to an interest in low impact, community living in the spirit of what would now be viewed as sustainability. Some despaired of urban living, and advocated instead a return to rural living, such as that illustrated in Figure 5.4. The central part of this sketch formed the cover of the radical technology magazine *Undercurrents* in 1977, with the full picture providing the centre spread. It is clear that the artist regarded urban life as something of a nightmare, but it must be interpreted in its contemporary setting (after the oil crisis and just before the 'Winter of Discontent'). Nevertheless, even at that time, this vision of an ideal, rural existence (freed from unwanted 'side effects' of modern living) was not without its critics even amongst the radical or alternative technologists. Figure 5.5 shows a graphic response that was published in the subsequent issue of *Undercurrents*: 'welcome to the real world OK'. 'Autonomy' was viewed as utopian in the 1970s and 'sustainable cities', its counterpart at the dawn of the twenty-first century, can be similarly criticized as impractical now.

Notwithstanding the above criticisms of 'autonomy', cities can be viewed as forming part of a broader sustainable community, stretching beyond the city boundary and drawing resources from its rural hinterland or 'bioregion'. This bioregional thinking attempts to emphasize the interdependence of cities and their natural surroundings. Berg (1990) argued that in order for cities to become more sustainable, they should secure a reciprocal dependence between their urban settlement and the surrounding bioregion. However, at current rates of consumption, the footprint of cities far exceeds their natural catchment. The least restrictive interpretation of a sustainable community would be one that is both resource efficient and relies only on products of sustainable production. A 'compact city' (Breheny, 1995) would have a major role in ensuring the most efficient use of resources within a given urban land area. It would enhance the energy efficiency of the building stock, promote the use of public transport and alternatives, such as walking and cycling, and improve the feasibility of waste recycling and reuse at common locations. But it would still depend, in large measure, on resources from beyond its physical boundary.

Environmental footprint analysis

Ecological or environmental footprints

The use of environmental footprint analysis has grown in popularity over recent years, both in Europe and North America. It provides a simple, but often graphic, measure of the environmental impact of human activity: whether in the foreseeable future we will be able to 'tread softly on the Earth' (Hammond, 2000a and 2000b). Resource use and wastes produced by a defined population are converted to a common basis: the area of productive land and aquatic ecosystems sequestered (in hectares) from whatever source in global terms. Its roots lie in earlier ideas, such as 'Ghost Acres', and similar concepts developed by Borgstrom (1972) and Ehrlich (1968) in the late 1960s. Rees used footprint analysis in its basic form to teach planning students for some 20 years (Wackernagel and Rees, 1996).

The terms 'environmental' and 'ecological' footprints are used interchangeably here, although the former is preferred. Ecology is that branch of biology dealing with the interaction of organisms and their surroundings. 'Human ecology', sometimes used for the study of humans and their environment, is closer to the usage implied by footprint analysis.

Methodology

Footprint calculations involve several steps. Initially the per capita land area appropriated for each major category of consumption (aa_i) is determined:

$$aa_i = \frac{c_i}{p_i} \sim \frac{\text{annual consumption of an item}}{\text{average annual yield}}, \frac{\text{kg/capita}}{\text{kg/ha}} \quad (5.1)$$

In the version of footprint analysis employed by Wackernagel and Rees (1996) four consumption categories are identified: energy use, the built environment (the land area covered by a settlement and its connection infrastructure), food, and forestry products. This is a restricted subset of all goods and services consumed, which was determined by the practical requirements of data gathering and influenced by the development of the technique in a Canadian setting. Nevertheless, unconnected work by Friends of the Earth Europe (1995) using the related 'Environmental Space' concept, adopted a similar set of categories. In order to calculate the per capita footprint (ef) used in the present work, the appropriated land area for each consumption category is then summed to yield:

$$ef = \sum_{i=1}^{i=n} aa_i \quad (5.2)$$

The calculation leads to a matrix of consumption categories and land use requirements, which is ideally suited to a spreadsheet implementation.

Critical assessment

The environmental footprint provides a quantitative basis for evaluating the environmental impact of a population and a means of raising awareness of the consequences of human activity. It is a valuable technique in a toolkit of measures that can aid the assessment of sustainable development; balancing economic and social development with environmental protection (Hammond, 1998, 2000a,b). Satterthwaite (1999) has devised a set of criteria for sustainability, including health and sanitation, recreational facilities and numerous other aspects of social provision. Clearly environmental footprint analysis would need to be supplemented by the use of other measures to account for these broader aspects of human welfare.

Footprint analysis implies judgements about the relative weighting of the various consumption categories and their environmental impact. It reduces all such impacts to a common basis in terms of hectares per capita, which may not prove to be a unit that can be readily assimilated by ordinary people. The International Institute for Environment and Development (1996) described the process of analysis whereby all environmental impacts are aggregated into a simple index as 'resource reductionism'. They likened it to traditional measures of economic welfare, such as the Gross Domestic Product (GDP); see Hammond (2000b). Nevertheless, it provides a useful basis for contrasting the footprint of human activity with the available land area. The consequences of human consumption can be graphically viewed against the 'carrying capacity' of a nation, region or the planet as a whole.

Case study: the city of Bath

The Georgian city

Bath is situated in the South West of England, between the Mendip Hills and the Cotswolds, and has a population of about 84 200. The architecture of the city centre is predominantly of the Palladian style (named after the Italian architect, Andria Palladio), built mainly in the period 1714–1830 when a succession of Georges (I to IV) reigned over the United Kingdom. The era is consequently known as 'Georgian'. Building in Bath really took off from 1726 when the river between Bath and Bristol was made navigable, and building materials could be imported into the city by water from Bristol. The characteristic soft, mellow (Oolitic) limestone was extracted from quarries owned by Ralph Allen (1693–1764) on nearby Combe Down. The city expanded dramatically from the original medieval core (depicted in Figure 5.2) to meet the needs of visitors, with new public spaces linked by terraced houses in the Palladian style. The old battlement walls were dismantled, and the farmland immediately surrounding the core was built upon, whilst most of the grander houses depicted by Gilmore (and in Figure 5.2) were demolished. Much of Bath's present architectural elegance is associated with John Wood (1704–54) and his son, both architects and developers, who planned Queen Square, the King's Circus, and the Royal Crescent. These were formally laid out in the old Manor of Walcot, a patchwork of small fields to the north and west of the medieval city walls (Woodward, 1992). 'Georgian Bath' remains the focus of the city's heritage and its world renown.

The supposed 'healing powers' of the Bath's spa waters were identified well before Roman times and have been used through the centuries (see pp. 87–9 above, and Woodward, 1992). However, one of the main reasons for the city developing rapidly in the eighteenth century was the visit by Queen Anne in 1702, followed by the aristocracy of the country. People came to Bath for a variety of reasons (both commercial and recreational) during that period, only one of which was to take advantage of the healing powers of the spa waters. The city was officially recognized by UNESCO in 1987, and is one of some 19 World Heritage Sites in Britain. Tourism is a major industry and a large proportion of the city's income and employment comes from this sector. The amount of available space for transport within the city centre is extremely restricted. Traffic congestion has become so severe in recent years that several feasibility studies have been undertaken to investigate the potential for pedestrianizing core areas. The income of Bath residents is generally higher than the UK average, although there are areas of relative disadvantage (such as Foxhill and Twerton).

Footprint analysis

The city of Bath is used here as a focus for an examination of the sustainability of cities in their wider context. A dataset was compiled in an analogous manner to that devised by Wackernagel and Rees (1996) for their footprint studies, principally in a Canadian setting. It reflects the novel features of Bath and its neighbouring bioregion (see also Doughty and Hammond, 1997 and 2000). The environmental space available on a city scale is roughly equivalent to the area of the watershed in which the city is located.

Figure 5.6 Widening of the system boundaries for environmental footprint analysis (Bath, Avon, the South West Region, and the United Kingdom)

Source: Doughty and Hammond (2000).

Table 5.1 Land use matrix and environmental footprint for Bath (ha per capita)

	Fossil energy	Built-up area	Arable land	Pasture	Forest	Sea	Total
Food	**0.02**		**0.21**	**0.70**	**0.01**	?	**0.94**
vegetarian	?		0.20				0.20
animal products	?			0.70		?	0.70
water							
Housing and furniture	**0.70**	**0.03**			**0.25**		**0.98**
Transport	**0.39**	**0.02**			**0.04**		**0.45**
road	0.30						0.30
rail	0.01						0.01
air	0.06						0.06
coastal and waterways	0.01						0.01
Goods	**0.70**		**0.01**	**0.22**	**0.18**		**1.11**
paper production	0.06				0.18		0.24
clothes			0.01	0.22			0.23
tobacco							0.00
others	0.64				0.20		0.84
Total	**1.81**	**0.05**	**0.22**	**0.92**	**0.48**	?	**3.48**

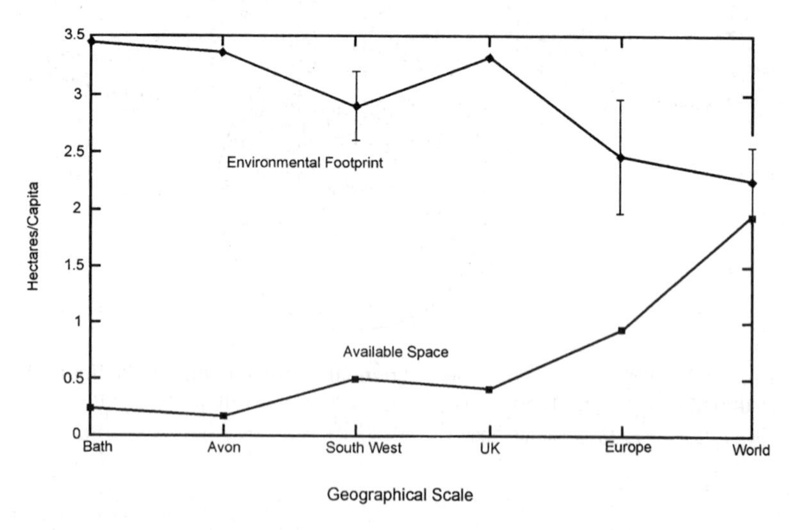

Figure 5.7 Geographic representation of environmental footprints
Source: Doughty and Hammond (1997).

In the case of Bath, the watershed of the former County of Avon has been divided here between its population, and then the appropriate amount allocated to the population of Bath city itself. This method avoids the problem of attempting to assign a natural bioregion to a city that shares its watershed with other populations, such as Bristol and Bradford on Avon. The available areas at the county, regional and other scales were defined by their geopolitical boundaries, and are illustrated in Figure 5.6 (Doughty and Hammond, 2000). The per capita land use matrix for Bath is presented in Table 5.1 (adapted from Doughty and Hammond, 1997). It is evident that the largest contribution to the overall environment impact is due to 'fossil energy', followed by 'pasture' and then 'forest'. The other components are significantly smaller, although the uncertainties in some of the components must be borne in mind. Bath's corresponding environmental footprint and that for surrounding bioregions is shown graphically in Figure 5.7. Doughty and Hammond (1997) argued that the point at which the environmental footprint curve intersects the available 'ecospace', indicates when enough land area is available to support a population sustainably. In the case of Bath and wider geographic areas, the footprint line does not cross the available area line, suggesting that even at the continental scale the sustainable carrying capacity has already been exceeded. This supports previous research conducted by Friends of the Earth (1995) and Wackernagel and Rees (1996), which found that most western lifestyles, such as those in Europe and North America, have consumption patterns that result in footprints which are far greater than the amount of geographically available land.

Sustainability and cities

An extensive literature on 'sustainable cities' has developed in the 1990s. One of the leading thinkers and advocates has been the so-called 'cultural ecologist', Herbert Girardet (1992 and 1999). Although he didn't initially use the term, *The Gaia Atlas of Cities* (Girardet, 1992) was one of the earliest texts to stimulate an interest in the role of cities as a major source of environmental damage: 'the city as parasite'. There he posed the question as to whether or not cities are sustainable, but did not really answer it. Instead he identified the many disbenefits of modern cities, and argued for a change in the way that they are planned and organized. Girardet noted that the inputs and outputs of urban living are unsustainable: finite energy resources and other material inputs with waste outputs. This he termed the 'linear metabolism' of cities. A more desirable system would be one that he called 'circular metabolism' in which the inputs are efficiently harnessed and the waste products are reduced, reused or recycled. Lord (Richard) Rogers of Riverside, the eminent modernist architect, has been influenced by Girardet's idea on sustainable cities. In the book based on his 1995 Reith Lectures (co-written with Philip Gumuchdjian, an Associate Director in his Partnership) he uses the notion of 'sustainable cities' and Girardet's concept of 'circular metabolism'. However, the former was given a broad definition (which might better be denoted by the expression 'convivial cities') that could encompass the views of many, disparate protagonists:

- *A just city*, where justice, food, shelter, education, health and hope are fairly distributed and where all people participate in government
- *A beautiful city*, where art, architecture and landscape spark the imagination and move the spirit
- *A creative city*, where open-mindedness and experimentation mobilize the full potential of its human resources and allows a fast response to change
- *An ecological city*, which minimizes its ecological impact, where landscape and the built form are balanced and where buildings and infrastructures are safe and resource efficient
- *A city of easy contact and mobility*, where information is exchanged both face-to-face and electronically
- *A compact and polycentric city*, which protects the countryside, focuses and integrates communities within neighbourhoods and maximizes proximity
- *A diverse city*, where a broad range of overlapping activities create animation, inspiration and foster a vital public life.

Richard Rogers also advocates 'sustainable urban planning', which he contends should involve citizens in decision-making at every level. [Something that should perhaps have been taken by the local authorities and planners to be a prerequisite; given the Latin root of 'civilization' alluded to in the Introduction.] However, by the time that Rogers chaired the UK Government's Urban Task Force (1999) the idea of sustainable cities had formally disappeared, although component parts of the broader concept of sustainable development remained centre stage. The Task Force was given the remit of determining an appropriate strategy for providing 4 million additional homes in England over the next 25 years. They recommended greater reuse of 'brown field' sites to develop new compact, cohesive settlements. Ironically, Girardet's support for the notion of 'sustainable cities' seemed to be solidifying (1999), although he now uses Rogers' broad definition reproduced above. This contains many desirable elements of a modern urban community, but they don't amount to a verification of the concept. In order to secure sustainability in the wider context, the greater inhibition is arguably the urban–rural divide. It is this interface or boundary over which most of the input resources for cities must pass.

Concluding remarks

Three-quarters of the world's population may live in cities by the year 2025 (Rogers, 1997). They consequently form a very important element of the human condition. It has been argued here that the notion of 'sustainable cities' is simply based on a misconceived idea of the full implications of sustainability, as well as the way that cities have developed historically. The technique of environmental footprint analysis can be employed to examine the sustainability of cities by placing them in their broader geographic context. The 'Georgian' city of Bath was adopted in the present work as a case study following the early research of Doughty and Hammond (1997 and 2000). Its per capita footprint was found to be greater than that of the surrounding bioregion, and indeed of the wider geographic areas. The environmental footprint of the city is nearly twenty times larger than that of the corresponding land area. This lends support to the authors' critique of the idea of sustainable cities popularized in contemporary literature. Cities only survive because they are inextricably linked by human, material and communications networks to their hinterland or bioregions.

Notwithstanding the above criticism of the idea of 'sustainable cities', urban design of compact cities can obviously contribute a more sustainable way of life, particularly in industrialized societies. This can be done by encouraging the development of integrated mixed-use urban communities in much the same way that has been advocated by a diverse range of architectural critics and urban planners. A key element in this type of development is to focus on greatly improving the efficiency of resource use within cities, and thereby reducing their environmental footprint. This will clearly enhance 'sustainability', although it is impractical to achieve the very strict system conditions laid down under 'The Natural Step'.

It is desirable to create over time cohesive and convivial cities that are diverse, yet socially balanced, communities in an attractive setting (Urban Task Force, 1999). This requires a conscious effort to reverse the trends in urban planning evident during most of the twentieth century. Sustainability assessment techniques need to be employed across the urban–rural interface in an extended process. Environmental footprint analysis could form an important part of that assessment.

Towards an environmental research agenda

The UK Government is currently in the midst of the second phase of its (Technology) Fore*sight* Programme aimed at generating 'visions for the future' that might assist decision-making in both the public and private sectors of the economy. This should provide a basis for ensuring that resources are used in support of wealth creation and improving the quality of life (Hammond, 2000a). The programme was launched in 1993 and now has 13 independent panels. Each panel consisted of members drawn from a wide range of academics, industrial and research backgrounds. They developed networks of business people, engineers and scientists to engage in the *Foresight* process. Individual panels operated in slightly different ways within an overall framework, although all seek to identify high priority technological projects that could be implemented (successfully introduced to the market) within a 20-year, medium-term time horizon. Nevertheless, this does not stop the panels 'visioning' over a longer time span, and examining the more speculative constraints and possibilities. The panels most closely associated with the present work are those covering 'Built Environment & Transport' (BET) and 'Energy & Natural Environment' (ENE). Following a recent national review it is intended to move away from a structure based on these standing panels over a five-year period.

The ENE *Foresight* Panel identified the rapid growth of cities (particularly in South East Asia and Latin America) as a major challenge for sustainability (Fore*sight*, 2000). They recommended R&D activities in the areas of less resource intensive city construction and infrastructure (the so-called 'Factor 10' improvements in productivity; see Hammond, 2000b), distributed utility systems and integrated transport. The development of low carbon technologies is an underlying theme in each of these areas. In addition, the Panel advocated 'closed-loop' production and consumption, with less hazardous materials, reduced waste, health and biodiversity damage. This is essentially Girardet's 'circular metabolism' concept, whose implementation and effectiveness can be assessed using environmental footprint analysis. The BET *Foresight* Panel's Construction Associate Programme have proposed the promotion of 'smart' buildings and infrastructure to create new business opportunities, improve living/working environments, and enable information feedback to improve construction quality. They recommend accelerating the introduction of new technologies, 'intelligent' products, standardized pre-assembled components and advanced materials into every level of the built environment. This 'high tech' approach has also been advocated in Girardet's more recent work (1999). Traditionally the construction industry has been perceived as being rather 'conservative'. The BET Panel therefore suggests that it should be encouraged to 'embrace sustainability' by adopting sustainable construction methods and whole life principles. They contend that by shifting the industry's culture towards sustainable thinking at every level, it 'can save energy, reduce waste and pollution and cut the lifetime costs of property ownership' (*Foresight*, 2001).

Acknowledgements

The research reported here forms part of a study supported by a research grant from the UK Engineering and Physical Sciences Research Council, under its 'Sustainable Cities' programme, entitled 'Monitoring the City' (GR/J92910, awarded to Professor Hammond jointly with Professor A.K. Day of the Department of Architecture and Civil Engineering, and Professor A. Lewis of the Department of Psychology). Professor Hammond would also like to acknowledge the support of British Gas plc (now demerged as BG plc and Centrica plc), who has partially funded his Chair. However, the views expressed in the paper are those of the authors alone, and do not necessarily reflect the policies of the original or the successor companies. Both authors wish to thank Mathis Wackernagel for the provision of an unpublished paper on the application of footprint analysis for the city of Santiago de Chile, and a corresponding spreadsheet program. Councillor Roger Symonds of B&NES Council made helpful comments on an earlier version of this work, particularly in relation to the historical development of the city of Bath. Finally, they are grateful for the care with which Sarah Fuge prepared the typescript and Gill Green prepared the figures.

References

Amin, S. (1997) 'Economic, political and social distortions in the modern world'. In P. Smith and A. Tenner (eds), *Dimensions of Sustainability*, pp. 19–25. Baden-Baden: Nomos Verlagsgesellschaft.

Barnett, J. (1987) *The Elusive City: Five Centuries of Design, Ambition and Miscalculation*. London: The Herbert Press.

Berg, P. (1990) *Growing a Life-Place Politics*. In V. Andruss et al. (eds), *Home! A Bioregional Reader*. Philadelphia: New Society.

Borgstrom, G. (1972) *The Hungry Planet: The Modern World at the Edge of Famine* (2nd edn). New York: Macmillan.
Breheny, M. (1995) 'The compact city and transport consumption'. *Trans. Inst. British Geographers*, 20(1), 81–101.
Bronowski, J. (1973) *The Ascent of Man*. London: BBC Publications.
Day, A.K. & Hammond, G.P. (1996) *Sustainable Cities*. International Centre for the Environment Seminar Series: 'Towards an environmental research agenda', University of Bath (unpublished), 18 June.
Department of the Environment, Transport and the Regions (1999) *A Better Quality of Life: A Strategy of Sustainable Development for the UK*. London: TSO.
Doughty, M. & Hammond, G. (1997) 'The use of environmental footprints to evaluate the sustainability of cities'. In N. Russel et al. (eds), *Proc 6th IRNES Conf: Technology, the Environment and Us*, pp. 170–6. London: Imperial College.
Doughty, M. & Hammond, G. (2000) 'Cities, environmental footprints and sustainability'. *City Development Strategies*, Issue 4, 40–3, Third Quarter.
Ehrlich, P. (1968) *The Population Bomb*. New York: Ballantine.
Ekins, P. & Cooper, I. (1993) *Cities and Sustainability*. Clean Technology Unit.
Foresight (2000) *Stepping Stones to Sustainability*. London: DTI.
Foresight (2001) *Constructing the Future*. London: DTI.
Friends of the Earth (1995) *Towards Sustainable Europe*. Brussels: FOE.
Girardet, H. (1992) *The Gaia Atlas of Cities: New Directions for Sustainable Urban Living*. London: Gaia Books.
Girardet, H. (1999) *Creating Sustainable Cities*. Totnes: Green Books/The Schumacher Society.
Hall, P. (1998) *Cities in Civilisation*. London: Weidenfeld & Nicolson.
Hammond, G.P. (1998) 'Alternative energy strategies for the United Kingdom revisited: market competition and sustainability'. *Technological Forecasting and Social Change*, 59, 131–51.
Hammond, G.P. (2000a) 'Energy and the environment'. In A. Warhurst (ed.), *Towards a Collaborative Environment Research Agenda: Challenges for Business and Society*, pp. 139–78. Basingstoke: Macmillan Press.
Hammond, G.P. (2000b) 'Energy, environment and sustainable development: a UK perspective'. *Trans. IChemE Part B: Process Safety and Environmental Protection*, 78, 304–23.
Hammond, G.P. (2001) *Engineering sustainability: Energy systems, heat transfer processes and thermodynamics analysis*. Invited Lecture: 7th UK National Heat Transfer Conference, Nottingham University, 11–12 September 2001, 32 pp.
Harper, P. (1976) 'Autonomy'. In G. Boyle & P. Harper (eds), *Radical Technology*, pp. 134–67. Wildwood House, London.
Haughton, G. & Hunter, C. (1994) *Sustainable Cities*. London: Jessica Kingsley Publishers.
International Institute for Environment and Development (1996) *Citizen Action to Lighten Britain's Ecological Footprints*. A report from the IIED to the UK Department of the Environment, IIED, London.
Parker, G. (ed.) (1986) *The World: An Illustrated History*. London: Times Books.
Porritt, J. (2000) *Playing Safe: Science and the Environment*. London: Thames & Hudson.
Rogers, R. (1997) *Cities for a Small Planet*. London: Faber & Faber.

Satterthwaite, D. (ed.) (1999) *The Earthscan Reader in Sustainable Cities*. London: Earthscan.
Thring, M.W. (1990) 'Engineering in a stable world'. *Science, Technology and Development*, 8(2), 107–21.
UK Round Table on Sustainable Development (1999) *Fourth Annual Report*. London: DETR.
Urban Task Force (1999) *Towards an Urban Renaissance*. London: E. & F.N. Spon.
Woodward, M. (1992) *The Building of Bath*. Bath: The Building of Bath Museum.
Wackernagel, C. & Rees, W. (1996) *Our Ecological Footprint*. Philadelphia: New Society.
World Commission on Environment and Development (WCED) (1987) *Our Common Future*. Oxford: Oxford University Press.

Further reading

Barnett, J. (1987) *The Elusive City: Five Centuries of Design, Ambition and Miscalculation*. London: The Herbert Press.
Boyle, G. & Harper, P. (eds) (1976) *Radical Technology*. London: Wildwood House.
Girardet, H. (1992) *The Gaia Atlas of Cities: New Directions for Sustainable Urban Living*. London: Gaia Books.
Girardet, H. (1999) *Creating Sustainable Cities*. Totnes: Green Books/The Schumacher Society.
Hall, P. (1998) *Cities in Civilisation*. London: Weidenfeld & Nicolson.
Haughton, G. & Hunter, C. (1994) *Sustainable Cities*. London: Jessica Kingsley.
Rogers, R. (1997) *Cities for a Small Planet*. London: Faber & Faber.
Rogers, R. & Power, A. (2000) *Cities for a Small Country*. London: Faber & Faber.
Russell, N., Byron, H., Dixon, A. & Richardson, J. (eds) (1997) *Technology, the Environment and Us: Proc. Sixth IRNES Conf.* London: Imperial College Graduate School of the Environment.
Wackernagel, M. & Rees, W. (1995) *Our Ecological Footprint: Reducing Human Impact on the Earth*. Philadelphia: New Society Publishers.

6

Financial Drivers of Environmental Performance: The Political Economy of Globalization and Liberalization in the Extractive Industries

Adrian Winnett

Summary

This paper:

- surveys and synthesizes some of the economic literature relevant to the international operations of enterprises in the extractive industries, especially the metals mining industries;
- from this develops a framework to identify the possible financial and environmental impacts of globalization and liberalization on the structure, conduct and performance of those industries.

The first step is necessary since there is not available at present a coherent economic analysis of international trade and investment in natural resources. The relevant analysis is partially addressed by a number of pieces of economic analysis which only partially overlap, some of which were developed some time ago and others of which have been evolving rapidly (as part of the increasing returns/imperfect competition 'revolution' in international economics, for example, Helpman and Krugman, 1985; Grossman, 1992). We cannot begin to hope, in the present paper, to construct such a coherent analysis, but we can provide some pointers that help in identifying the likely financial and environmental effects of increasing international integration of natural resource markets. In fact, such an analysis must push beyond the bounds of the conventionally economic since it involves states and political groupings within states as essential actors, hence the use of the term '*political* economy'.

Globalization and liberalization: a thumbnail sketch

Globalization is usually taken to refer to a qualitatively new form of international political economy and its associated social and cultural characteristics. Different commentators use the term more or less widely and with various interpretations. There is considerable controversy not only about the meaning and usefulness of the term but also about the existence of the phenomena to which it supposedly refers. There is, though, agreement that an essential characteristic of globalization is the emergence of 'footloose' capital that is not identified with particular nation states. Much of the controversy is about whether, in fact, the world has seen the emergence of such footloose capital (for example, Hirst and Thompson, 1996).

Globalization can be regarded as a further stage in the evolution of various types and degrees of *internationalization*, which refers to the integration of international markets in outputs of goods and services and in productive inputs, especially of financial resources and of knowledge, including managerial skills (for example, Michie and Grieve Smith, 1995). As will be shown, many of the issues with which we are concerned do not require us to become embroiled in conceptual arguments about globalization. Much of what is most important simply flows from the observation of increasing levels of international market integration.

Market integration can be measured along two dimensions: the *volume* of transactions, and the penetration and *speed of adjustment* of markets: in particular, the speed with which markets arbitrage away perceived differences in net returns. Indices of integration based on the volume of transactions may be less significant than is often supposed. For many of the issues with which we are concerned, processes of market adjustment may be the more significant aspect.

Liberalization is closely related to the enhancement of market adjustment processes, attempting to increase the coverage of markets and the efficiency with which they work. It typically has three main components:

- *privatization* – the transfer of public sector enterprises and agencies to the private sector;
- *deregulation* – the removal or relaxation of government controls over private sector economic activity, coupled with a preference for the use of market-based instruments if there is a need for policy intervention, for example, to correct for environmental externalities;
- *market* benchmarking – residual public sector activities should meet standard economic criteria of efficient operation.

A major issue in appraising liberalization programmes is the extent to which gains in economic efficiency, both static and dynamic, have been secured at the expense of increased inequity in income distribution. From the point of view of the present paper, an important question is the impact of capital imports, and especially foreign direct investment (FDI), on the recipient country's income distribution. This has always been a sensitive issue, and liberalization has given discussion of it a renewed impetus.

This anticipates the point that in an international context, liberalization is closely linked to attempts to increase the openness of economies, that is the proportion of national output involving international transactions (measured by internationally traded output and by productive activities involving overseas stakeholders). Negative barriers to trade and factor mobility may be removed or relaxed. These may be explicit barriers such as restrictions on currency convertibility, or hidden barriers such as bureaucratic difficulties in obtaining licences or undisclosed preferential purchasing and hiring practices. This leads into positive encouragement of openness through the provision of predictable and enforceable legal frameworks for economic transactions and, more broadly, by an assurance of good governance. Alongside distributional issues, and related to them,is the important question of the extent to which there may be a tension between increased openness and growing demands for environmental protection.

In developing and transitional economies, liberalization is frequently linked to the provision of financial resources through IMF stabilization packages and/or longer term World Bank structural adjustment packages. (For a lively critical account of the Indian experience which also raises many general issues, see Bhaduri and Nayyar, 1996.)

Some basic economic issues

Most of the recent literature on globalization has been primarily concerned with the manufacturing and services sectors, and with horizontally integrated or conglomerate multinational enterprises (MNEs). (Horizontal integration is defined as the ownership of several 'plants' at the same stage of the productive process; conglomerates own essentially unrelated productive activities. MNEs own such plants or activities across national boundaries.) The primary commodity industries, though the prototype for many forms of internationalization, have been relatively neglected in the recent mainstream literature, as has the vertically integrated MNE. (Vertical integration is defined as the ownership of plants at different upstream or downstream stages of the productive process: both are clearly relevant to the issues of this paper.)

Of course, much of the general discussion of internationalization and multinational enterprises is relevant to the extractive industries, but some is not and much of it needs refinement to deal with the characteristics of extractive industries, particularly those related to:

- the locational inflexibility and
- the time profile of resource extraction.

In relation to the first point some care is needed. It is easy to assert that the reason why much of the discussion of globalization is not relevant is because extractive industries are not 'footloose', as they are tied to their deposits. Leaving aside the general question of whether or not any industries are really footloose, this misses the point: ownership can be footloose even if productive activities are not (and there may, in fact, be considerable locational choice for new mining ventures anyway). The issues are therefore more subtle; part of the purpose of this paper is to draw attention to these subtleties.

There was much discussion of extractive industries in the aftermath of the 'oil shocks' of the 1970s and of the potential for similar cartelization in other primary commodities (*vide* the dates of some of the references provided to this paper). Even without considering changes in the global economic environment, it is not obvious how transferable much of this discussion really is outside the oil industry, and to the metals industries in particular. Nonetheless, it highlighted the nature and significance of bargaining over resource rents among income claimants at different stages of the activities through which extractive resources are transformed into final output. The scope for and the outcomes of bargaining are conditioned by locational inflexibility *and* have implications for the time pattern of extraction. Analysis of these bargaining processes is integral to any discussion of the related impacts of globalization and environmental pressures on the international mining industry.

The economic model of exhaustible resource depletion which was first laid out by Hotelling as long ago as 1931 (and in many essentials by Gray in 1913; see, for example, Hartwick and Olewiler, 1986, Chapter 3 for a standard textbook account) has recently become central to wider discussions of sustainability. The prototype sustainability question is this: if there is a finite resource which is essential to productive activity, can an economy sustain indefinitely some specified level of output? (Somewhat surprisingly, under restrictive assumptions the answer is 'yes'; see, for example, Hartwick and Olewiler *op. cit.*, Chapter 6.) This question can be refined in various ways, and the answer has important practical applications. A critical issue in formulating the answer is the level of investment of resource rents. In practice, this is presumably not unrelated to bargaining over rents and so is also not unrelated to the point made in the preceding paragraph.

We have laid out some preliminary points derived from the recent history of discussion of the international extractive resource industry. We now look at these in a little more detail and, then *via* a discussion of risk management, formulate a framework for investigating the interrelated impacts of increasing internationalization of financial markets and responsiveness to an environmental agenda.

First, though, this is a convenient point at which to provide a map through the main argument of the remainder of the paper:

(i) As is clear from the preceding discussion, one line of analysis has been concerned with MNEs – which from one point of view are really just a particular form of international investment. Many of the relevant efficiency and distributional issues can thus be analyzed in a quite general capital-mobility framework.
(ii) From another point of view MNEs can be seen as an alternative form of economic organisation to the market, using internal hierarchies rather than external markets to organize resource production.
(iii) In turn, extractive MNEs are a particular type of MNE characterized by the nature of their income-generating processes. The choice between these modes of organization is conditioned by the scope for bargaining over resource rents and this then influences the time pattern of resource production over time.
(iv) The costs of accepting or handling various types and levels of risk affect time patterns of extraction both directly and indirectly, through their impact on modes of organization.
(v) The rates of return used to evaluate programmes of resource production are crucial. Both risk and return are influenced, in more or less complex ways by increasing integration of and innovation in international financial markets *and* by the acceptance of increased environmental responsibilities by MNEs.
(vi) But to add further complexity, both the pressure and the ability to accept and handle environmental risk is also conditioned by changes in financial markets.

International capital mobility

In principle, capital should migrate internationally to equalize risk-adjusted marginal net returns to investment. This is not only wealth-maximizing for private transactors, but, on the improbable assumption of no divergence between private and social returns (such as those caused by environmental pollution), socially optimal for the countries involved and also globally.

However it does have income-distributional effects. Assuming (as is traditional in economics) diminishing returns to investment, if capital migrates from lower return to higher return countries, output and other factor incomes decline in the former, but are offset by higher incomes to capital owners, an increased amount of whose capital is now located overseas. In the latter, returns to domestic capital decline, but are more than offset by increases in the income of other factors. In this model, 'other factors' are usually thought of as labour, but may include the services of natural resources. Further, the labour force may be differentiated and this may be associated with, for example, employment opportunities in foreign-owned enclaves, often supposed to be a characteristic of primary commodity production. In these cases, the distributional effects among 'other factors' are sensitive to the specification of production functions and factor supplies. (This is mostly all standard textbook economics, for example, Brenton, Scott and Sinclair, 1997 – which is not to say that it does not provide important insights.)

Increasing internationalization and liberalization of markets should accelerate the process of return equalization. The wider, associated enhancement of market processes should also lead to convergence in risk premia. Thus observed net returns should converge across countries, since the returns required by and offered to investors should be less sensitive to particular local conditions.

The conventional wisdom is also that if capital markets are more allocatively efficient, enabling better matching of borrowing and lending, the total stock of capital should also rise and thus average returns to capital decline, at least in the long run. (Lenders can more easily achieve their desired lifetime profiles of consumption and risk-return mixes of assets. In the short run, though, some observed returns may rise as markets are deregulated.) Certainly, average risk premia should fall; this is independent of the effects on risk premia from financial market innovation to be discussed later.

Thus we would expect to find:

- more homogeneity in risk-adjusted returns on capital
- lower equilibrium risk-adjusted returns on capital

both of which have implications for the time profiles of natural resource extraction programmes.

Anticipating our later discussion, the first point means that we would expect, *ceteris paribus*, the planned horizon of extraction programmes to become more similar across locations, or, to put it another way, to become relatively more sensitive to differences in labour and other extraction costs, which may include the costs of meeting environmental obligations.

The second point means that planned extraction horizons should lengthen which does not necessarily mean that total extraction will fall, since lower capital costs may mean that more programmes are started, if there is a fixed cost element in investment – for example from exploration and construction expenditures, as there undoubtedly is. In short, we would anticipate more and longer-lived mining investment. Thus we cannot, say, in any simple way, that the industry becomes more sustainable, if by this is meant that total extraction declines: horizons may lengthen, but scale may increase. However, we should also note that if environmental obligations impose heavier shut-down and post-closure expenditures, these will weigh relatively more heavily if projects are evaluated at lower returns.

Further, there may be macroeconomic as well as industry-level effects. If the global capital stock is higher, so are average incomes. If, as in many recent discussions, there are so-called endogenous growth (that is, essentially, if there are not diminishing returns to capital) effects, incomes may not only be higher but may also grow faster. Thus demand for natural resources may be higher than it otherwise would be, increasing prices and extraction rates in the long run.

Private investors seek to equalize returns net of taxes (and other deductions). Given the advantages of higher capital stocks, governments therefore have an incentive to reduce tax rates on internationally mobile capital assets. The equilibrium outcome is, in fact, zero taxation of mobile capital (or indeed, other mobile factors). With increasing integration of international capital markets which both prevents holding capital captive and places countries in competition for inward investment, revenue-constrained governments will tend to shift taxation to immobile factors. This is usually thought of as labour, and some of the resentment of foreign investment flows from this distributional effect.

However, natural resources are also immobile. If taxation is appropriately designed, that is, if only the resource rent element is taxed, incentives to extract from established operations should not be affected (see, for example, Hartwick and Olewiler *op. cit.*, pp. 68–70; for a detailed analysis of one case, Gillis, Beals et al., 1980). If the extraction programme maximized the owners discounted wealth before tax, then it would obviously do so after. There are, though, a number of complications:

- in practice it may often be difficult to design taxes which are neutral in this way;
- in any case, incentives to start-up new operations would be affected by international differences in resource taxation regimes and, in addition, these may be complementary with capital inflows;
- if taxes are used to correct for environmental externalities and if these differ from country to country, then this creates complicated policy conflicts.

We can regard the tax regime as a proxy for a variety of possible bargains between foreign investors in natural resources and host governments, and maybe other claimants on incomes from those resources, and can thus use this as a framework for analyzing possible outcomes in terms of extraction and exploration patterns.

The multinational enterprise

MNEs are vehicles for FDI, and thus much of the preceding discussion of capital flows is applicable. However, packaged with the capital, MNEs provide information, especially managerial and technological. The major question in understanding the role of MNEs is to establish the cost advantages of internalizing transactions within an organization rather than trading across markets, given that there is a presumption that costs of transactions will be higher across national boundaries due to differences in legal systems, language, social norms, and so on. (A standard survey is Caves, 1982.) The accepted framework for analyzing these cost advantages is based on the work of Williamson (1975) on horizontal and vertical integration. Essentially, they derive from the need to deal with information asymmetries, small-numbers opportunism, and sunk costs in transactions.

This can be easily understood by thinking of a mineral industry example of transactions between extractors and processors. Suppose there are large numbers of buyers and sellers, and good information about qualities and availability of supplies. There is no reason why continuous spot transactions would not be the lowest form cost of organization. If there is a lack of information about qualities or availability, there will be an incentive to form long-term contracts to handle these uncertainties. These, however, can be complex and costly to negotiate, monitor and enforce. If there are sunk costs in organizing contracts, this in effect produces a small numbers situation, since even if there are notionally large numbers of buyers and sellers, it will be costly to switch partners.

This explains the incentives to vertical integration (Williamson *op. cit.*, Chapter 5; Caves *op. cit.*, Chapter 1.2; Casson et al., 1986, Part 1). In fact, the extent of vertical integration varies widely in the non-ferrous metals industry, both across sectors and over time. (For historical, but analytically relevant, accounts: aluminium, Stuckey, 1983; copper and tin, case studies in Casson et al. *op. cit.*) We would expect the costs and benefits to be finely balanced, since, along with most primary commodities, the metals industries have long-established, sophisticated grading systems and market-based futures and options contracts. However, the existence of these systems and contracts itself points to the significance of the types of features which can drive vertical integration. Taking environmental considerations into account adds to informational complexity, since, in effect, not only the product but also the production process has to be monitored.

There is also considerable horizontal integration in the industry. Essentially the same explanatory framework applies. However, it is possible to think of a simple prototype (Helpman and Krugman, 1985). There are economies of scale in what are now commonly called headquarters services and the location of production is distributed according to relative costs. This is just a particular application of the so-called new trade theory which emphasizes economies of scale effects from concentrating certain sorts of activities. It is readily applicable to the mining industries. Here we would expect significant headquarters economies from the informational requirements not only of production and markets but also, especially of exploration, and we would expect considerable relative cost variations across space and time not only from labour and other input costs but more specifically from geological conditions. If environmental considerations become important we would expect that this will increase the headquarter's scale advantages, from collecting, evaluating and monitoring environmental information on a comparative basis, and add an additional factor into the relative costs of different production sites.

We can thus provisionally conclude that taking into account environmental considerations probably leads to increased relative cost advantages from both vertical and horizontal integration. Given that MNEs have to have cost advantages from integration that outweigh the assumed additional costs from operating across national boundaries, this enhances those advantages unless environmental information is highly localized and difficult to transfer.

Internationalization and liberalization have features which impact on the costs of MNEs.

- Greater social and cultural homogeneity erode cross-boundary cost differentials, as does adherence to predictable and enforceable standards for contracts. This is a dynamic process, since MNEs are themselves a major vehicle for bringing about internationalization and liberalization.
- On the other hand, increased financial integration and innovation may remove some cost advantages which depend on having privileged access to particular sources of finance.

The foregoing picture needs complicating in at least two ways.

- MNEs in the metals industries generally exhibit both vertical and horizontal integration: the main effect (and probably cause) of this is to influence the interrelated resource-rent bargains struck with ultimate resource owners, with non-integrated upstream or downstream enterprises, and with parallel firms in the industry.
- We have presented internal organization and market contracts (whether spot or long-term) as the relevant alternatives, but there may be various hybrids, especially joint ventures, which add further transaction cost comparisons into the picture and the relative advantages of which may shift as a result of growing environmental pressures and international integration.

Natural resource rents

Capital invested in extractive processes should earn a competitive risk-adjusted rate of return. To the extent that there are elements of monopoly or impediments to capital mobility, rates of return will diverge from this. However, exhaustible resources will also generate an element of rent or royalty reflecting their ultimate scarcity: that is, a unit of resource extracted now is necessarily a unit less extracted later.

Under some simple assumptions the extent of this is easily derived. Suppose there is a known, homogenous stock of some resource, the marginal extraction costs of which do not vary with the rate of extraction, then we can reason as follows:

- from the point of view of a profit-maximizing producer, a unit of resource extracted at any date must be worth the same as that extracted at any other date, and thus the rent on the marginal unit extracted now must be the same as the discounted rent on the marginal unit extracted later;
- alternatively, from the point of view of an investor, a unit of resource held 'in the ground' must earn the same risk-adjusted return as some

alternative investment and since the return is the increase in the value of the resource stock, the rent on the marginal unextracted unit of resource must rise at the rate of return.

Either way the rent on a unit of resource must rise at the rate of interest (that is, the discount rate used by the resource producer or the rate of return used by the resource investor). This is the well-known 'Hotelling rule'. This can be interpreted descriptively and prescriptively. It applies to a perfectly competitive resource extracting industry with perfect foresight. Such an industry should reduce its rate of extraction over time so that the price of a unit of resource rises at the interest rate. Alternatively, if there are no externalities (such as environmental pollution) in either the production or the consumption of the resource, and if the interest rate is socially appropriate, this is also the socially efficient rate of extraction.

As pointed out above, if processes of internationalization and liberalization lead to more convergence of rates of return and to lower rates of return, both because underlying rates converge and fall, and because risk premia converge and fall, we would expect extraction horizons to converge and to lengthen. (The implicit assumption is that with restricted capital mobility, individual metal-producing industries are themselves, to some extent, segmented. As far as we know, there is no discussion in the literature of resource extraction paths in segmented industries, but it would seem important for understanding the impact of internationalization and globalization.)

The Hotelling rule can be extended in various ways. For our purposes, two points can be noted:

- Suppose we consider an individual extractor ('mine'), selling on a market which offers a given price and whose marginal extraction costs rises with the rate of extraction, then they should follow the same rule, achieved by reducing extraction rates over time so as to lower marginal costs (this was Gray's earlier version of the argument). In practice we would expect to find both processes at work, with industry output being redistributed among mines appropriately.
- If there is some degree of monopoly in the industry, then firms will take account of the impact that their extraction rates have on market prices. Provided that demand for the resource becomes more elastic as it becomes scarcer (which is not unreasonable: greater price sensitivity at higher prices), monopolistic firms will adopt longer time horizons since the perceived marginal pay-off to reducing extraction rates is greater than for competitive industries.

The Hotelling rule appears to be a simple testable proposition. The numerous attempts to test it have been, at best, inconclusive. It is important, though, not to lose sight of the basic proposition that exhaustible resource industries should exhibit some scarcity rent, even if the 'rent-rising-at-interest-rate' version seems not to be observed. The reasons for the empirical 'failure' of the rule are suggestive of questions which are important for the present paper.

Firstly, it is difficult to identify the scarcity rent separately from other returns, 'normal' and monopolistic. Thus, in less precise policy contexts, it is easy to see why resource taxes, or generally bargaining over claims on resource rents, are distortionary in their impact on production decisions even if this is unintended.

Uncertainty and bargaining

However, the major difficulty in extending the rule is to cover situations of uncertainty. This needs spelling out in more detail.

There may be industry-level uncertainty about:

- available reserves or the outcome of exploration activities;
- market conditions, either because of 'imperfect foresight' (lack of forward markets), or because of strategic behaviour by producers in relation to extraction or exploration, or because of speculation in already extracted stocks.

Clearly these sources of uncertainty are:

- interrelated and, to some extent, reinforcing;
- empirically all highly significant in the non-ferrous metals industry;
- basic to the explanation of the existence and scope of MNEs, for reasons discussed earlier.

To properly analyze the workings of the industry there is a need to bring together the analyzes of industry structure and of patterns of resource extraction rents over time. But to a considerable degree, the standard models of these are based on inconsistent sets of assumptions, and integrated models are rather special and piecemeal. In addition there may be sources of uncertainty outside the industry, which may interact, of course, with those in the industry.

Anticipating our later discussion: to the extent that firms assume environmental responsibilities, this adds in a further risk factor. However, this may be offset by reductions in risk elsewhere if firm and market efficiency is improved by globalization and liberalization, and by increased capacity to handle risk through financial innovation (although all these may themselves be partly offset by some increase in associated systemic risk).

Resource extractors have to strike bargains with ultimate resource owners, which in developing countries generally means the state. Both parties have an interest in efficient programmes of exploration and extraction. Whereas the latter are, in principle, relatively well defined, the former are not. Exploration is an uncertainty-reducing activity and thus there is an incentive to renegotiate bargains as information is revealed. This renegotiation may be difficult to avoid due to the prospector's sunk costs. The foreknowledge of this influences the initial bargains that are struck, and these initial bargains may change over time as negotiators gain more knowledge both of the outcome of prospectors' activities and of each other. This process is sometimes referred to as the 'obscolescing bargain'. Part of the objective of liberalization is to remove the threat of arbitrary renegotiation, through assuring security of property rights and enforceable contracts.

Further, there is complementarity between exploration and extraction. This is partly direct: extraction yields information about the size and quality of already discovered reserves and information about geological conditions which assists in locating further reserves. It is also indirect: the resource extractor and owner gain information about each other which may be used in subsequent negotiations.

To the extent that environmental obligations add an extra cost into exploratory activity, not only through the regulation of the activity itself but also, more importantly, through the need to more fully assess possible sites, then the sunk cost element and scope for renegotiation increases. Thus it becomes more important to protect against arbitrary renegotiation.

Similar arguments apply to the extraction phase itself: there is an incentive to take advantage of the extractor's sunk costs in order to renegotiate bargains. This can shift extraction away from mutually beneficial time paths. This is reinforced by the difficulty of identifying what might be called the 'bargainable' resource rent from the going risk-adjusted rate of return. Again environmental obligations raise sunk costs and therefore the scope for renegotiation, and this may be mitigated by liberalization strategies which remove expropriation and similar risks. Additionally, more integration and efficiency in international capital markets may help to better identify required rates of return.

If resource acquiring firms do not fully integrate upstream into prospecting and extraction, but organize or partly organize resource supplies through the market, then bargaining over resource rents occurs if there are monopoly positions. Most interest has focused on the ability of resource sellers, particularly state-controlled enterprises, to form and maintain cartels. (For a sophisticated analysis, typical of the work inspired by OPEC, especially, see Kemp and Van Long, 1984.) For well-known reasons, cartels tend to be unstable: basically there is always an incentive to increase output at the price set by the cartel. Internationalization and liberalization probably undermine the position of cartels still further. (United Nations, 1987 provides a standard check list of the factors affecting the bargaining strength of firms and governments.)

The general conclusion is that internationalization and liberalization probably reduce the short run share of resource owners in individual resource rent bargains (or more broadly, in resource revenues), a share which acceptance of environmental obligations had the potential to increase through sunk cost effects, but in the longer run we would expect more bargains in total and possibly some increase in the shares if more firms compete for resource rights.

It is important to distinguish these share effects from long-run price trends dependent on the shifting of demand functions by factors such as technical innovation in substitutes and recycling. Some of this is itself driven by environmental pressures. One of the difficulties has been that share and price effects have not always been properly distinguished, and resource owners have attempted to protect against price effects by bidding up their shares of resource rents, with by now familiar implications for the efficiency of resource discovery and extraction.

Markets in risk

MNEs have been well-placed to take advantage of international financial market imperfections since they have low costs of arbitrage. They have also been able to offer diversification: holding their equity proxies for international portfolio diversification (Caves *op. cit.*, Chapter 6). With increasing integration and efficiency in international financial markets, these advantages will presumably become less apparent.

In markets for metals, sellers and buyers along with those for other primary commodities have long used a sophisticated array of derivatives to protect against price risk. (For gold, see Tufano, 1996, which gives a general account of the instruments available.) Indeed derivatives originated in these markets. They have now been widely adopted in financial markets. In principle they should be efficiency-enhancing, but there is a widespread perception that they can aggravate market instability, in the presence of imperfections in information and/or irrational transactors (World Bank). From the point of view of individual firms, financial market innovation should enable them to reduce the costs of managing certain sorts of risk, such as currency risk, but this may be partly offset if the innovations themselves aggravate systemic risk, as is often supposed. More generally, to the extent that innovation enhances the market efficiency which comes with more integration, it should lower expected returns, partly directly through declines in risk premia, partly indirectly through lowering underlying rates of return as market liquidity improves and so on. Again though, we should be alert to the effects of possible increases in systemic instability

As environmental risk becomes more widely accepted as a responsibility of firms in the industry, financial markets will increasingly develop contracts to deal with it, perhaps with some support from national and international agencies. This process is already well under way. As the market develops, we would expect to see more accurate assessment of risks and liabilities, and therefore more efficient contracts.

Firms essentially have three strategies in relation to risk: reduction, management, and offsetting.

There is a perhaps sanguine belief that reductions in environmental risk can come about at little cost as a result of moving to best practice and by improvements in best practice. The argument is that there are synergies between environmental improvement and efficiency improvement (for example, Auty, 1995, pp. 203–5, based on work with Warhurst). If firms are required to reduce risks, we can then calculate the residual costs, and the resulting impact on production plans. If they are required instead or as well as to accept liabilities, we can then ask what the cost minimizing mix of risk reduction and risk management, through insurance, is, and then assess the impact on production plans.

We can think of firms as accepting some costs of risk. If the costs of some components fall, they may then take on additional costs elsewhere. Costs may fall either because the events become less likely or costly or because there are cheaper methods for handling them, through risk reduction or management. (For a standard discussion, see Herring, 1983.) As a result of internationalization and liberalization, we would expect to see both processes at work for some non-environmental risks. For example, there is now a greater capacity to deal with currency risk and less risk of expropriation. Thus it may be easier for firms to handle increased environmental risks. As against this, there may be increased socio-political risks from the shifts in income distribution associated with liberalization and from systemic financial instability (*vide* South-East Asia). Part of reason for the increase in environmental awareness is that local populations increasingly treat environmental degradation as a component of their living standards: thus failure to reduce risks may itself be interpreted as an adverse distributional shift.

Conclusion: towards a political economy research agenda

In the preceding discussion we have already identified what appear to be the major positive effects of international integration and liberalization of financial markets:

- convergence in risk-adjusted returns;
- probably some lowering of expected risk-adjusted returns;
- increased capacity in against certain kinds of risk, leading to convergence and lowering of risk premia.

We would expect these to lead to more uniformity and longer time horizons in production programmes in extractive industries such as non-ferrous metals mining. This is a sustainability gain not in the simple sense that the horizon is longer, but in the sense that it is probably closer to what is socially efficient. Against this is increased incentive to start up more projects. In terms of micro-level effects, the outcome depends not only on the balance between *horizon* effects and *start-up* effects, but also on how environmental externalities are related to *scale* of operation as compared with *length* of operation. Finally, given that many externalities in mining are post-closure, using lower discount rates will weight these more heavily.

Thus globalization and liberalization have complex indirect effects through financial markets on the extent of environmental risk and on the sustainable use of resources. But they also have direct effects on the extent of environmental risk and on the system's capacity to handle it.

- Firms will find it easier to shift to locations with lower costs of all sorts and this includes the costs of meeting environmental standards. As with taxation, there is a risk of competitive laxity in standards to attract mobile capital.
- But there is pressure towards more uniformity of standards, partly as a deliberate international strategy (in response, partly, to the first point), and partly as a result of convergence across countries of expectations about environmental performance as a component of living standards, though this process is slow and erratic.
- We have discussed above the relationship between changes in financial markets and environmental handling capacity. Again the conclusions are not unambiguous.

The impact on the horizons of extraction plans of higher environmental costs is not clear cut. If they simply apply like an increase in extraction costs over the lifetime of the operation, they will simply extend the lifetime of the operation: current prices will rise, but future prices will be lower than they otherwise would be. However, again the issue is not just one of horizons, but of whether they are more or less efficient: taking into account externalities should move them in the right direction.

Some of the interactions between finance and the environment are mediated through their effects on organizational structure, the role and functioning of MNEs, and, closely related, the bargaining processes over resource rents. MNEs are presumably in a good position to implement environmental management, since it is probably an activity with significant economies of scale, but environmental commitments could worsen their bargaining position because of the sunk costs involved. However, the underlying processes of internationalization and globalization, should lead to negotiations over resource rents which have more pre-commitment. (As always, only detailed empirical work will illuminate these issues: for example, we are currently researching, at the University of Bath, the design and implementation of contracts in the Indonesian mining industry.)

This leads once more into the wider issue of sustainability. A benchmark for sustainability is that depletion of resources should be efficiently managed (essentially following the Hotelling rule) and that resource rents should be invested, not consumed. One of the major difficulties for many resource-rich countries has been that resource rents have been consumed and that their economies have thus performed badly in the long run (Auty *op. cit.*, Part 5; Heady, McGregor and Winnett, 1997, Chapter 3), as well as increasing their vulnerability to short-run shocks in resource markets. Another ingredient of this 'resource–curse' argument is that resource-rich locations are prey to internal and external plunder and thus to international and civil strife, dramatically shortening the time horizons of resource extractors. Is the basic sustainability criterion more likely to be met if resource owners take a smaller share of the rent, as seems possible with globalization and liberalization? This is largely a question of how international capital markets operate under the new global dispensation. There are some grounds for optimism that resources may become better managed in terms both of extraction paths and of environmental externalities, but will the rents be successfully recycled into investment?

References

Auty, R. (1995) *Patterns of Development*. London: Edward Arnold.
Bhaduri, A. & Nayyar, D. (1996) *The Intelligent Person's Guide to Liberalization*. New Delhi: Penguin.
Brenton, M., Scott, H. & Sinclair, P. (1997) *International Trade*. Oxford: OUP.
Casson, C. et al. (1986) *Multinationals and World Trade*. London: Allen & Unwin.
Caves, R. (1982) *Multinational Enterprise and Economic Analysis*. Cambridge, UK: CUP.
Gillis, M. et al. (1980) *Tax and Investment Policies for Hard Minerals*. Cambridge, Mass.: Ballinger.
Grossman, G. (1992) *Imperfect Competition and International Trade*. Cambridge, Mass.: MIT.
Hartwick, J. & Olewiler, N. (1986) *The Economics of Natural Resource Use*. New York: Harper & Row.
Heady, C., McGregor, J. & Winnett, A. (1997) *Poverty and Natural Resources* (typescript).
Helpman, E. & Krugman, P. (1985) *Market Structure and Foreign Trade*. Cambridge, Mass.: MIT.
Herring, R. (ed.) (1983) *Managing International Risk*. Cambridge, UK: CUP.
Hirst, P. & Thompson, G. (1996) *Globalization in Question*. Cambridge, UK: Polity.
Kemp, M. & Van Long, N. (1984) 'The role of natural resources in trade models'. In R. Jones & P. Kenen (eds), *Handbook of International Economics Volume 1*. Amsterdam: North Holland.
Michie, J. & Grieve Smith, J. (eds) (1995) *Managing the Global Economy*. Oxford: OUP.
Mining Journal. *Emerging Markets* (annual report).
Stuckey, J. (1983) *Vertical Integration and Joint Ventures in the Aluminium Industry*. Cambridge, Mass.: HUP.
Tufano, P. (1996) 'Who manages risk? An empirical examination of risk management practices in the gold mining industry'. *Journal of Finance*, 1097–137.
United Nations Centre on Transnational Corporations (1987) *Transnational Corporations and Non-Fuel Primary Commodities in Developing Countries*. New York: UN.
Williamson, O. (1975) *Markets and Hierarchies*. New York: Free Press.
World Bank. *Global Development Finance* (annual). Washington: World Bank.

References

Part IV

Valuing the Environment

7
When is a Spade not (only) a Spade? When it's an Environmental Management Tool

Eamonn Molloy

Summary

This paper shows how Life-Cycle Assessment (LCA), an environmental management technique, is constructed and mobilized in relation to a diverse range of interests and agendas that extend beyond the assessment of environmental impacts of products and processes. The paper argues that this political dimension of LCA is largely obscured by the dominant framing of the technique as 'scientific', 'objective' and 'value-free'. Characterization of LCA in this way, the paper suggests, is the result of deep-seated epistemological commitments of LCA professionals on the one hand and a desire to promote and establish the technique within business and policy cultures on the other. This argument is supported by an analysis of three texts that seek to promote LCA to different audiences. Through this analysis, the paper draws attention to the way that the rhetoric of 'science', 'objectivity' and 'value-neutrality' functions as a political resource. The principal implication arising from this finding is that the particular kinds of politics embodied within LCA need to be understood if use of the technique is to contribute positively to improved environmental performance.

Introduction

The starting point of this paper is that the formal definitions of LCA obscure a great deal about what the technique is actually able to accomplish, how the technique is employed and the diversity of reasons for its use in different situations. The fact that each step of the methodology (goal definition, inventory analysis, impact assessment) requires practitioners to use their professional judgement is usually not addressed by definitions of LCA which identify the process as value-free, objective, and transparent. In addition, the technique is employed by industry, policy makers and NGOs for a variety of purposes that extend beyond the assessment of environmental impacts to include marketing strategies, improving competitiveness, local and global policy formulation, challenging policy instruments and so on. In this paper, I want to address the questions of how the clearly political, social and economic dimensions of the technique are hidden in the definitions of it and how this is achieved. I argue that LCA has been constructed as scientific, objective and value free partly as a result of the epistemological commitments of LCA professionals *and* partly in order to establish the technique as a legitimate policy instrument within contemporary business–policy culture. In order to make this argument, I first of all present my assumptions about the characteristics of this business–policy environment and then provide a reading of three different texts to show how the LCA technique is framed around these characteristics. These three texts have been selected because they are widely recognized as key sources of information on LCA and because they are directed at different audiences.

LCA in policy context

Profound social and technological changes in the late-twentieth century has been accompanied by a widespread awareness of and concern with the anthropogenic risks associated with these developments. Earlier societies faced risks such as famine, diseases and drought that were 'natural' hazards in the sense that they were not produced directly as the result of human involvement with the bio-physical world. Modernity promised liberation from this sort of risk through the application of 'scientific' knowledge to the project of controlling nature. However, the enormous power of scientific methods coupled with (and constituted by) an increasing objectification and disregard for nature supported by utilitarian, instrumentalist economic 'rationality', not only failed to reduce the number of risks encountered in everyday life but actually led to their proliferation. In fact, modernity has successfully produced risks of catastrophic potential for the entire planet.

As the ideologies of market relationships, consumer 'choice', and economic rationality pervades ever more aspects of our lives, it is increasingly obvious that the 'market' circulates 'bads' as well as 'goods' and that the risks that threaten our well-being are the product of human choices (Simmons, 1996). With this recognition comes the anxiety that the activities of consumer society may not be sustainable and may already have exceeded 'the limits' (Meadows, Meadows and Randers, 1971).

The anxiety mentioned above is the focus of a growing body of scholarship that points to the sociological significance of risk and risk management in 'high modernity' (see, for example, Giddens, 1990; Luhmann, 1993). Beck (1992) suggests that risk has become such a defining feature of contemporary society that it is possible to speak of a 'risk society'. Defining the nature, scale and distribution of risk – for example, environmental, health or social risks – has become a central arena of struggle between diverse collective actors (Abel, 1985; Lau, 1992). These new agonistic relations escape representation in terms of traditional conflict structures or institutionalization processes which makes coherent collective action more difficult for those involved (Lau, 1992). These arenas of debate also re-figure official representations of the relationship between scientific knowledge and public policy:

> the more scientific findings and arguments are used as strategic resources the more the idea of a technical and instrumental use of objective and definite scientific results becomes obsolete or even proves to be a social fiction. (Lau, 1992, p. 243)

The emergence of widespread public environmental concern since the late 1960s can be seen as an expression of the anxiety arising from the culturally pervasive consciousness of risk (Giddens, 1990). As arenas of collective action, environmental issues are characterized by shifting fields of conflict, that constantly form and reform, and within which scientific knowledge is mobilized increasingly to strategic ends (Simmons, 1996). The specific issues, technologies and products around which these anxieties and conflicts coalesce may be quite adventitious, shaped as much by social factors as by the existence of 'objective' risks; for example, by the opportunity structure created by institutional arenas such as public enquiries, by arena characteristics that render some issues inherently more 'winnable' for environmental NGOs, or by the occurrence of specific incidents that provide a focus for public concerns and collective action (Grove-White, 1993). Similarly, what counts as 'green' or 'environmental' is the result of different degrees of negotiation, argument and involvement. Whose definition gets to count, where, when and for whom is a political and moral question that cannot be resolved by the realist, deterministic assumptions of science and policy institutions alone (Haraway, 1997).

The implicit social and cultural dimensions of environmental issues introduce ambiguities and tensions that are largely avoided by policy makers as a result of the realist framework within which they operate. These aspects of environmental debates also limit management capacities to predict or control the ways in which they are perceived by publics and consequently their level of credibility with these constituencies too. Recognizing LCA as shaped by and contributing to the formation of these frameworks is important because it illuminates how the technique is configured by the particular politics and values associated with them.

Guidelines for Life-Cycle Assessment: A 'Code of Practice' (SETAC, 1993a)

This text contains the most frequently cited definition of LCA. Perhaps more than any other organization, SETAC has shaped the contemporary debates surrounding LCA identity and methodology through organizing conferences, debates, workshops and the publication of these guidelines. This document is regarded as an authoritative key text among LCA practitioners despite various modest assertions by SETAC that it is not a prescriptive standard and is offered only as a 'guideline' (SETAC, 1993a, p. iii). The definition is as follows:

> Life-Cycle Assessment is a process to evaluate the environmental burdens associated with a product, process or activity by identifying and quantifying energy and materials used and wastes released to the environment; to assess the impact of those energy and material uses and releases to the environment; and to identify and evaluate opportunities to affect environmental improvements. The assessment includes the entire life cycle of the product, process or activity, encompassing extracting and processing raw materials; manufacturing, transportation and distribution; use, re-use, maintenance; re-cycling and final disposal. (SETAC, 1993a, p. 5)

SETAC clarify this definition further by stating:

> LCA addresses environmental impacts of the system under study in the areas of ecological health, human health and resource depletion. It does not address economic considerations or social effects.

A number of important implications arise from this definition. First, the insistence that LCA *does not* address economic or social issues can be interpreted as an attempt to effect closure on the debate as to the scope of LCA. Second, the distinction clearly indicates the assumption of the authors that 'environmental' (technical) issues can meaningfully be separated from socio-political issues. Exactly how 'ecological health, human health and resource depletion' can be distinguished from economic or 'social' issues is not explained by SETAC. The assumption that nature and culture are distinct elucidates SETAC's positivist epistemological commitments. This commitment to a nature/culture divide is also present in SETAC's list of 'the prime objectives of carrying out an LCA' (SETAC, 1993a, p. 5). These are:

(i) to provide as complete a picture as possible of the interactions of an activity with the environment;
(ii) to contribute to the understanding of the overall and interdependent nature of the environmental consequences of human activities; and
(iii) to provide decision makers with information which defines the environmental effects of these activities and identifies opportunities for environmental improvements.

Immediately following this list of 'prime objectives' (and notably unnumbered) SETAC suggest: 'the systematic procedures for LCA facilitate constructive dialogue among different sectors in society concerned with environmental quality'.

In addition to highlighting their view of the environment as an entity 'out there' that humans interact *with* and are independent *of*, this list of objectives constructs the 'different sectors of society' in a particular manner. For instance, SETAC suggest that 'constructive' dialogue is 'facilitated' (perhaps used instead of 'determined') by 'the systematic procedures for LCA.' So, in this view, not only is a systematic approach the most effective way to understand objective nature (the environment), it is also useful for facilitating understanding on the other side of the divide – that is, within the social world. The suggestion here is that constructive dialogue is systematic, and dialogue that isn't systematic isn't constructive. This little rhetorical device invites the reader to either buy into SETAC LCA methodology and engage in constructive dialogue or remain silent.

However, the document itself points to the difficulty in maintaining the society/nature boundary by arguing that LCA facilitates a symmetrical treatment of each. More worryingly, if we take the origin of the society/nature divide to be rooted in a modernist distinction between (social) subject and (natural or technological) object (Latour, 1993) society and the human subject become objectified. It is perhaps because of this working model of an objectified human subject deficient in systematically produced and delivered information, that SETAC seriously undervalues the role of wider consultation on LCA. In this view, LCA scientists and technical experts mediate between nature and society by virtue of their ability to transform knowledge of one into information for the other. The information deficient public has no role other than as passive recipient of monologic information and, as such, will not be part of any constructive dialogue. Consequently, mention of public involvement in SETAC's LCA literature constitutes little more than a 'tip of the hat' to wider debates about public participation.

A second point arising from these objectives is that SETAC see one of the functions of LCA as 'to provide decision makers with information'. Unlike 'the public', decision makers' information needs qualify for classification under 'prime objective' status, maintaining the putative one-way information flow from scientists' objective, realist account of nature to decision makers' subjective and ideological worlds.

However, there is a difficulty here. While SETAC attempt to systematically objectify the social, they also construct 'the social' as sufficiently different from (objectified) nature to require the continued mediation that (their) scientific knowledge can provide in the form of LCA. Seen in this light, what this means is that the objectification of nature has been accomplished but the project of objectifying the social is ongoing and necessary. The construction of LCA as an unproblematic technique is instrumental in achieving this goal. Nonetheless, by explicitly mobilizing LCA in relation to these social goals, the political dimension of LCA is revealed. As an organization with an explicit interest in promoting and stabilizing the technique, SETAC would clearly wish to avoid this situation. Here, the rhetoric of scientific objectivity and neutrality serves to construct the technique as outside the realm of 'politics' (values) and consequently above problematization in these terms.

Guidelines for the application of LCA in the EU eco-labelling programme

The *Groupe des Sages* was set up by the European Commission in December 1993 to advise on the role of LCA in the EU eco-labelling programme. The *Groupe des Sages* consisted of six prominent LCA experts, one of whom was also a member of the UK eco-labelling board, another was President of the Society for the Promotion of Life-Cycle Development (SPOLD), and a third edited the SETAC 'Code of Practice' on LCA. This group produced the document 'Guidelines for the Application of Life-Cycle Assessment in the EU Eco-Labelling Programme' (*Groupe des Sages*, 1994). In this document there is little acknowledgement of the uncertainties associated with scientific knowledge production, but instead a concern that these processes should remain distinct from 'politics'. The stated objectives of the guidelines were:

(i) to support the Member States in establishing eco-labelling criteria which are based on a methodology which is scientifically sound and workable in practice, and
(ii) to improve uniformity in the methods applied in different Member States.

At least partially in response to the above objectives, the 'Guidelines' strongly frame LCA in technocratic terms. The definition of LCA used in the guidelines comes from the SETAC 'Code of Practice' (SETAC, 1993a) and is presented as an authoritative definition without the reminder that it is not intended as a prescriptive standard. Throughout the document, LCA methodology is presented as unproblematic and uncontroversial. Serious methodological issues that are the subject of considerable debate within the LCA technical literature, for example establishing system boundaries, are raised only as '*points of discussion*' leading to '*research needs*'. In other words, LCA is presented as a well-established technique. The description of LCA as 'scientifically sound', and able to 'improve uniformity' can be interpreted as supporting the presentation of LCA as an appropriate instrument for realizing wider policy commitments of the European Commission such as harmonization, unification and standardization (Waterton and Wynne, 1996). Further, given the common professional commitments of the authors to establishing an LCA network, this translation of the European Commission represents an attempt to achieve official recognition of the technique. In other words, these authors mobilize LCA to establish a network in which LCA, and the organizations the authors represent, are essential to the EU eco-labelling scheme.

The list of objectives also suggests the model of science embodied in the document. This is a 'traditional' view which separates theory (abstract knowledge) from practice (the application of knowledge) (Barnes, 1982). This is exemplified by the first objective that is: 'to support the member states in establishing eco-labelling criteria which are both scientifically sound and workable in practice.' The use of a distinction between 'pure' and 'applied' knowledge here simultaneously maintains the privileged epistemological status of scientific knowledge and constructs this knowledge as useful for practical policy purposes. The authors of such abstract knowledge thus justify their supposed detached status and construct themselves as indispensable to society at the same time. The second objective 'to improve uniformity in the methods applied in different Member States' can be read as an assumption that science discovers universal truths, thereby facilitating standardization or 'uniformity'. Ultimately, this objective relies on the theory/practice distinction too, in that current practices (applied methods) are seen – problematically – to vary across member states because of the absence of adequate universal truth (theory). The basic idea is that insofar as abstract scientific knowledge is universalized, diversity of practice will be eliminated. However not only does this model fail to acknowledge the diversity of current LCA practice, it also fails to account for the diversity of practices that constitute and construct what counts as scientific knowledge itself.

These positivistic commitments are articulated further through the document's 'key recommendations'. These are:

(i) LCA can make a significant contribution in providing a scientific, unifying and transparent basis for the EU eco-labelling programme.
(ii) It is central to this programme because it compares different products on the basis of their common function.
(iii) It relates environmental impacts at all stages from cradle to grave to both market changes and technology improvements.
(iv) It is a methodology still in the process of development, requiring additional research and systematic data collection.
(v) Therefore policy makers, competent bodies and practitioners must remain aware of the current capabilities and limitations of LCA and should support its continuous development (*Groupe des Sages*, 199, p. 1).

Within the context of the EU eco-labelling scheme, LCA is constructed as something that always had all the features of a legitimate instrument for use in a realist policy environment. It is well defined, uncontroversial, scientific, rational, objective, systematic and transparent. In addition, although it is presented as a technical 'tool' (therefore, in this view 'apolitical'), it is useful for relating environmental impacts to wider political goals, in particular market changes, technological development, comparability, standardization and 'harmonization'. However, before the promises of LCA can be fulfilled, there is a 'need' for more research which policy makers and others 'must' be aware of and 'should' continue to support. This last observation/recommendation represents a directly political prescription which is ostensibly to do with cautious adoption of LCA in the eco-labelling programme but which may also be seen as a loaded prescription for the EU to continue to support/resource the interests of the authors and, presumably the organizations they represent. Thus, the *Groupe des Sages* once again construct LCA as crucial to the EU eco-labelling scheme.

There are a number of effects resulting from this construction of LCA and the accompanying policy prescriptions. First, by constructing LCA as unproblematic, the *Groupe des Sages* effect a degree of technocratic closure on some important debates surrounding LCA methodology and application. The apparent agreement over LCA identity among the *Groupe des Sages* is not characteristic of LCA discourse in general and raises questions about the extent to which this group are representative of LCA practitioners and about who is qualified to count as an 'expert' in the view of the European Commission. Second, by constructing LCA as instrumental to wider EU political goals, the *Groupe des Sages* appear to be trying to achieve institutional acceptability for their version of LCA. However, this particular construction relies upon the ability of LCA to be transferred across sites which in turn rests upon the presentation of LCA as stable and uncontroversial. Third, as a result of successfully constructing LCA as suitable for use in the EU policy arena, the epistemological assumptions inherent within both the EU policy framework and this construction of LCA become predominant in shaping the direction and emphasis of the eco-labelling programme, so setting the terms for engagement with wider audiences. From here on out, participation in the eco-labelling debate must be couched in similar technical/technocratic terms (with associated values) if they are to be officially recognized as valid. Thus, other 'non-official' discourses are invalidated.

The LCA Sourcebook: A European Business Guide to LCA

'*The LCA Sourcebook: A European Business Guide to LCA*' was produced in 1993 as a joint venture between SustainAbility [sic], Society for the Promotion of LCA Development (SPOLD) and Business in the Environment (BiE). This document does not claim to be a comprehensive technical guide to LCA methodology, instead, the aim is to:

> help introduce the methods, applications and implications of LCA to encourage greater use by companies of the growing number of European centres of LCA excellence. (p. 8)

As such this text explicitly aims to enrol new actors into an LCA network. The style of the document is typical of 'green business' literature in that the language is strongly evangelical and proffers both managerialist prescriptions and technical academic analysis. In addition, this literature is informed by a unitary view of organizations where lack of environmental commitment is seen as the result of the wrong organizational culture or technology deficit (Newton and Harte, 1997). The LCA Sourcebook seeks to remedy this by demonstrating that LCA is the 'right way forward'.

The document begins with forewords by SPOLD (self-identified as 'A Catalyst for LCA Development') and BiE (self-identified as 'Practical Tools for Business'). These forewords set out to persuade the reader of a pressing need for businesses to 'adopt LCA' by referring to 'the environmental challenge' and the *need* for 'sustainable development'. LCA is variously described as a 'tool' for businesses that is 'emerging', 'objective', 'transparent', 'commonly accepted' and with 'scientific integrity.' The forewords also develop a futurist sense of urgency with statements like 'At this stage of development of LCA *your* opinion can be most influential', or 'The LCA Sourcebook is as useful to seasoned LCA practitioners as to those who have not *yet* put their foot on the ladder' (emphasis added). This theme is recurrent throughout the entire document, with LCA constantly referred to as 'evolving', 'emerging', 'expanding', 'embryonic'. Thus, LCA is constructed as a *promising* technology (Van den Belt and Rip, 1997). Further, the reader is (rather vaguely) reminded of the 'need to move *towards* LCA' as if LCA was a destination at a fixed time and place in a certain, unquestionable, uni-directional future. This prescription is bolstered by prolific reference to the 'burgeoning literature on the subject [LCA]', the 'growing number of European centres of LCA excellence' and a rhetorically impressive list of major corporations that are already involved with LCA (Volvo, Bosch, Henkel, Rhone-Poulenc, Smith-Kline Beecham).

To support this vision of LCA as 'the right step forward', a particular historical narrative is constructed in the Sourcebook. The following extract exemplifies this:

> Until very recently, however, life-cycle assessment (LCA) was of little interest to anyone outside of a small community of scientists, mostly based in Europe or North America. But then their work escaped from the laboratory and into the real world. Indeed, many of these early LCA pioneers have been astonished to see LCA emerging as a priority issue on both the regulatory and business agendas. (p. 8)

This narrative clearly illustrates a number of points about the models of science, LCA and business with which the authors are operating. First, the authors locate the origins of LCA with scientists in laboratories as an attempt to legitimize LCA on the grounds of its scientific heritage. This point may at first glance seem to be contradicted by the second sentence, which notes that 'their work escaped from the laboratory and into the real world'. This could be taken to imply that the knowledge produced in laboratories is so abstract as to be meaningless in grounded reality. However, this is not necessarily the case in that the construction of the laboratory (and by implication scientists and their practice) as outside of the 'real world' draws upon a particular 'traditional' view of science and scientific knowledge as detached, impartial, and beyond the realm of both the 'natural' and 'social' worlds (Barnes, 1982). This supposed scientific 'disinterestedness' is in part what enables scientists to claim value-neutrality and objectivity for the knowledge they produce. As a result, if LCA can be constructed as having this kind of heritage then it should be more appealing to business managers who operate in an economic and political climate where recourse to supposedly value-neutral knowledge strategically serves to free decision makers from accountability and responsibility for judgements based upon it (Porter, 1995).

Second, the use of the expression '*their work* [LCA] escaped *from*' raises some issues around the degree of agency attributed to LCA itself. Bearing in mind the considerable disagreement that presently exists over what LCA actually is, or should be, then what is the 'it' that 'escaped'? There is an implication here that LCA is black-boxed (Latour, 1987), to the extent that it is easily recognizable as such, and also at a stage of development (maturity) where it is ready for application in the 'real world', in which it will be useful to a host of different people. Further, given that scientists are supposed to operate free from economic, political or moral obligation in order to produce 'objective' knowledge, why did 'it' have to 'escape' rather than wait to be disseminated without prejudice across society? It may be that this notion of LCA 'escaping' from the laboratory is a way of articulating the positivist linear science–policy model that actively distinguishes between the 'world' of science (nature) and the 'world' of policy (society). The intention here may be to suggest that LCA is no longer in the hypothetical prison of science but has jumped the fence to the 'real world' of business and economics to become a valid technology for managers.

Another interpretation of this 'escape from the laboratory' metaphor is that it projects the inherent usefulness and practicality of LCA 'real-world' applications in contrast to its laboratory life where it simply constituted 'pure' research. This distinction between 'pure' and 'applied' is indicative of a traditional understanding of science and technology where abstract, objective and theoretical scientific knowledge is utilized in grounded, practical technology. What this model amounts to is the over-idealized distinction that science is about the discovery of truth whereas technology is about the application of truth (Barnes, 1982). Scholars from within the field of Science and Technology Studies (STS) have shown that science emerges from prior applications as experiments, for example, nuclear waste clean-up.

Finally, from the above, it would hardly be surprising that 'many of the early LCA pioneers have been astonished to see LCA emerging as a priority issue on both the regulatory and business agendas', if these scientist/pioneers really believed in a fixed, material boundary between science and policy. However, evidence of this astonishment is difficult to find elsewhere in LCA literature. Most accounts of the development of LCA explicitly locate the origins of the varied practices that may be thought of as LCA in companies that carried out the research for specific commercial purposes (Hunt and Franklin, 1996). Further, many of these accounts are written by early practitioners who have actively and successfully developed careers in LCA consultancy and promotion. However, it is possible to conceive that within the commercial organizations that developed LCA, scientific and technical departments were perceived (internally and externally) as detached from the rest of the organizational culture by virtue of the supposed unique nature of the work that they carried out and the knowledge that they produced. There seems to be no reason to assume that the model of scientific knowledge production as impartial, disinterested and unique should not operate within organizations at a departmental level as well as at the institutional level, especially as this model is the bedrock of 'modernity' *per se*.

In sum, this document attempts to persuade businesses of the value of LCA. The authors construct sustainable development and environmental responsibility as problems for business that need to be addressed, and LCA as the key response strategy. Further, LCA is presented as established, defined, transferable, stable yet also flexible. It is stable and defined in the sense that it is 'scientifically sound' and transferable in that it can easily be applied at diverse sites.

Discussion

Each of the above texts attempt to persuade particular audiences, for example LCA practitioners, the European Union, industry, of the relevance of LCA to their interests. The framings of LCA in each of the texts differ most in terms of the contextual anchor points that the authors use; for example, in the SETAC *'Code of Practice'*, LCA is framed largely in terms of standardization of technique; in the *Groupe des Sages* 'Guidelines', LCA is contextualized in terms of EU policy goals; and in the *LCA Sourcebook*, LCA is contextualized predominantly in terms of business strategy and economic objectives. This fluidity in terms of the framing of LCA highlights the way the texts at once address, problematize and construct their audiences and the authors. It is important to point out here that there is some degree of overlap in authorship of the three documents which at first may be obscured by the different orientation and style of the texts. For example, as mentioned earlier, some members of the *Groupe des Sages* edited the SETAC *'Code of Practice'* and others are members of SETAC. One author of the *Groupe des Sages* 'Guidelines' was also a member of the UK eco-labelling board. Similarly, one member of the *Groupe des Sages* was at the time the President of the Society for the Promotion of Life-cycle Development (SPOLD) and co-authored *'The LCA Sourcebook'* along with SustainAbility and Business in the Environment. This shared authorship is highly significant in accounting for the level of agreement between the texts to do with the deeper, tacit construction of LCA. All three refer to the scientific credentials of LCA, the SETAC definition, LCA's objectivity, neutrality, transparency, uniformity and its ability to facilitate constructive dialogue and communication.

The fact that the authors of the texts felt the need to legitimize LCA by invoking the rhetoric of science can be largely be accounted for by remembering that LCA has its epistemological roots within the 'natural' sciences, and has been developed within the same modernist institutional frameworks that it seeks to serve. As a result, it embodies many of the same values and commitments as its social, political, economic and cultural context. Ironically, by emphasizing LCA's scientific credentials and simultaneously using these as a rhetorical resource to make LCA appeal to policy makers and industry, the authors inadvertently demonstrate the political nature of the dichotomy between science and society that they seem to assume is only natural. The spheres of science and politics become indistinguishable.

The implicit constructions of 'nature' or 'the environment' embodied in LCA discourses are such that environmental problems can be defined exclusively in terms of objective physical processes. This assumption is problematic because just as environmental problems have their origins in particular patterns of social activity, so their meaning is socially negotiated. The 'environment' is not something external to society, but is implicated in the complex patterns of social and economic activity in which we engage. This is not to suggest that scientific practice and politics are the same, but that the constructions adopted by the promoters of LCA reflect a narrow technical definition of environment which attempts to maintain a clear distinction between spheres of value and does not problematize human relations and culture, for example consumerism as a 'way of life'. This particular definition of environment has been shaped by epistemological commitments that are shared, as a result of years of policy dialogue, by industry and government but which do not necessarily coincide with the broader moral and cultural constructions of environment that inform the views of large sections of the population. This may be one reason that LCA (especially in relation to eco-labels) has had little resonance or credibility with NGOs or wider publics.

An important point arising from this concerns the openness of environmental debates to public participation when the issues are framed solely in technical terms. The LCA texts examined in this research all suggested that LCA could facilitate 'public involvement' without describing how this would be done in practice. The emphasis within the texts on common, standardized, universalized language 'facilitated' by 'systematic' procedures delineates acceptable terms of engagement for policy debate. The resulting effect is that those groups who may wish to conduct dialogue in diverse discourses, or share a different understanding of 'nature' are excluded from the predominant monologue. The question is essentially whether debates about environmental issues should be conducted within a democratic or technocratic arena.

Conclusion

This paper set out to explore LCA as 'politics by other means'. It has shown that LCA is more than just a process to systematically evaluate environmental burdens associated with products or production processes. LCA responds and functions as a response to a range of different pressures and influences, including regulations, certification with voluntary schemes such as EMAS and the need to support marketing strategies. However, the fact that LCA is more than 'just' an environmental management tool is not lost on the professional organizations that seek to promote the technique. For example, SETAC promote LCA as a device that can '... facilitate constructive dialogue among different sectors of society'; the *Groupe des Sages* suggest it can contribute towards meeting the EU's goals of uniformity and harmonization of policy across member states; and SPOLD and BiE promote the technique on the grounds that it is 'good for business'. The question then arises as to why the definitions and constructions of the technique insist that the technique has no political, ethical or economic dimensions. In response, I argued that LCA is framed in this manner partly as a result of its epistemological heritage in the natural sciences, and partly as a promotional strategy in order to frame the technique as a legitimate tool for use in a realist business–policy culture.

References and further reading

Abel, R. (1985) 'Risk as an arena of struggle'. *University of Michigan Law Review*, 83(4), 712–812.

Akrich, M. (1992) 'The de-scription of technical objects'. In W. Bijker & J. Law (eds), *Shaping Technology/Building Society: Studies in Sociotechnical Change*. Cambridge, MA: MIT Press.

Akrich, M. & Latour, B. (1992) 'A summary of a convenient vocabulary for the semiotics of human and nonhuman assemblies' in W. Bijker & J. Law (eds), *Shaping Technology/Building Society: Studies in Sociotechnical Change*. Cambridge, MA: MIT Press.

Barley, S.R. & Kunda, G. (1992) 'Design and devotion: surges of rational and normative ideologies of control in managerial discourse'. *Administrative Science Quarterly*, 37(3), 363–99.

Barnes, B. (1982) 'The science–technology relationship: a model and a query'. *Social Studies of Science*, 12, 166–72.

Barnes, B. & Edge, D. (1982) *Science in Context: Readings in the Sociology of Science*. Milton Keynes: Open University Press.

Beck, U. (1992) *Risk Society: Towards a New Modernity*. London: Sage.

Bensahel, J. (1992) 'Assessing eco-impacts'. Paper presented at the 'Eco-labelling, Life Cycle Analysis and the Chemical Industry' conference, Brussels, 23–24 November.

Berkhout, F. (1996) 'Life cycle assessment and innovation in large firms'. *Business Strategy and the Environment*, 5, 145–55.

Berkhout, F. & Howes, R. (1997) 'The adoption of life-cycle approaches by industry: patterns and impacts'. *Resources, Conservation and Recycling*.

Bloomfield, B.P. & Best, A. (1992) 'Management-consultants: systems-development, power and the translation of problems'. *Sociological Review*, 40(3), 533–60.

Bloomfield, B.P. & Vurdubakis, T. (1994) 'Re-presenting technology: IT consultancy reports as rextual reality constructions'. *Sociology*, 28(2), 455–77.

Bloor, D. (1976) *Knowledge and Social Imagery*. London: Routledge & Kegan Paul.

Bowker, G. & Star, S.L. (1991) 'Situations vs. standards in long-term, wide-scale decision making: The Case of the International Classification of Diseases'. *Proceedings of the 24th Hawaiian International Conference on Systems Science*. Washington: IEEE Computer Society Press.

Braun, B. & Castree, N. (eds) (1998) *Remaking Reality: Nature at the Millennium*. London: Routledge.

Callon, M. (1986a) 'Some elements of a sociology of translation: domestication of the scallops and fishermen of St. Brieuc Bay'. In J. Law (ed.), *Power, Action and*

Belief: A New Sociology of Knowledge? Sociological Review Monograph. London: Routledge.
Callon, M. (1987) 'Society in the making: the study of technology as a tool for sociological analysis'. In W. Bijker, Hughes, T. & T. Pinch (eds), *The Social Construction of Technological Systems*. Cambridge, MA: MIT Press.
Callon, M. (1991) 'Techno-economic networks and irreversibility'. In J. Law (ed.), *A Sociology of Monsters: Essays on Power, Technology and Domination*. London: Routledge.
Callon, M. & Latour, B. (1981) 'Unscrewing the big leviathan: how actors macrostructure reality and how scientists help them to do so'. In K. Knorr-Cetina & A. Cicourcel (eds), *Advances in Social Theory and Methodology: Towards an Integration of Micro and Macro Sociologies*. London: Routledge.
Callon, M. & Law, J. (1982) 'On interests and their transformation: enrolment and counter-enrolment'. *Social Studies of Science*, 12, 615–25.
Cec (1992) *Towards sustainability: A European Community programme of policy and action in relation to the environment and sustainable development*. COM(92), 23, final. Brussels: Commission of the European Communities.
Clarke, A. & Montini, T. (1993) 'The many faces of RU486: tales of situated knowledges and technological contestations'. *Science, Technology and Human Values*, 18(1), Winter, 42–78.
Clift, R. (1993) 'Life cycle assessment and ecolabelling'. *Journal of Cleaner Production*, 1(3–4), 155–9.
Collins, H.M. & Pinch, T. (1993) *The Golem: What Everyone Should Know About Science*. Cambridge: Cambridge University Press.
Corino, C. 'Legal aspects of LCA: European legislation and jurisdiction'. *International Journal of Life Cycle Assessment*, 1(2), 66.
Curran, M.A. & Young, S. (1996) 'Report from the EPA conference on streamlining LCA'. *International Journal of Life Cycle Assessment*, 1(1), 57–60.
De Ooude, N. (1992) 'Establishing a common methodology'. Paper presented at the 'Eco-labelling, Life Cycle Analysis and the Chemical Industry', conference Brussels, 23–24 November.
EEC (1992) Council Regulation (EEC) No. 880/92 of 23 March 1992 on a Community eco-label award scheme. *Official Journal of the European Communities*, No. L 99, 11 April, pp. 1–6.
Elkington, J. (1993) 'The uses and abuses of life cycle analyses by non-governmental organisations'. *Journal of Cleaner Production*, 1(3–4), 151–3.
Fleck, L. (1935/1979) *Genesis and Development of a Scientific Fact*. Chicago: University of Chicago Press.
Fujimura, J.H. (1992) 'Crafting science: standardised packages, boundary objects and translation'. A. Pickering in (1992) *Science as Practice and Culture*, Chicago: Chicago University Press.
Giddens, A. (1990) *The Consequences of Modernity*. Cambridge. Polity.
Grahl, B. & Schmincke, E. (1996) 'Evaluation and decision-making processes in Life Cycle Assessment'. *International Journal of Life Cycle Assessment*, 1(1), 32–5.
Groupe des Sages (1994) 'Guidelines for the application of Life-Cycle Assessment in the EU ecolabelling programme'. Unpublished report to the European Commission.
Grove-White, R. (1993) 'Environmentalism: a new moral discourse for technological society?' In K. Milton (ed.), *Environmentalism: The View from Anthropology*. London: Routledge.

Guinee, J.B., Udo De Haes, H.A. & Huppes, G. (1993) 'Quantitative life cycle assessment of products, 1: goal definition and inventory'. *Journal of Cleaner Production*, 1(1), 3–13.
Haraway, D. (1988) 'Situated knowledges: the science question in feminism as a site of discourse on the privilege of partial perspective'. *Feminist Studies*, 14(3), 575–99.
Haraway, D. (1997) *Modest Witness*. London: Routledge.
Harding, S. (1992) *Whose Science? Whose Knowledge? Thinking from Women's Lives*. Ithaca: Cornell University Press.
Hindle, P. & De Oude, N.T. (1996) 'SPOLD – Society for the Promotion of Life-Cycle Development'. *International Journal of Life-Cycle Assessment*, 1(1), 55–6.
House of Commons (1991) Session 1990–91, Environment Committee, Eighth Report. *Eco-labelling*, vols. 1 & 2, HC 474-I & II, 24 July 1991. London: HMSO.
House of Commons (1993) Session 1992–3, Environment Committee. *Eco-labelling*, Minutes of Evidence, Wednesday 27 January 1993, UK Eco-Labelling Board; Department of the Environment; Department of Trade and Industry, HC 429-i. London: HMSO.
Huang, E.A. & Hunkeler, D.J. (1995) 'Life cycle concepts as management tools for minimizing environmental impacts'. *Center Report No. 131*, US–Japan Center for Technology Management, Vanderbilt University, Nashville, TE.
Hunt, R. & Franklin, W.E. (1996) 'LCA – how it came about'. *International Journal of Life Cycle Assessment*, 1(1), 4–7.
Jasanoff, S. (1990) *The Fifth Branch: Science Advisers as Policy Makers*. Cambridge, MA: Harvard University Press.
Jasanoff, S., Markle, G., Petersen, J. & Pinch, T. (eds) (1994) *Handbook of Science and Technology Studies*. London: Sage.
Knorr-Cetina, K.D. (1981) *The Manufacture of Knowledge: An Essay on the Constructivist and Contextual Nature of Science*. Oxford: Pergamon.
Kogut, B. & Parkinson, D. (1993) 'The diffusion of American organising principles in Europe'. In B. Kogut (ed.), *Country Competitiveness: Technology and the Organising of Work*. New York: Oxford University Press.
Kuhn, T. (1962) *The Structure of Scientific Revolutions*. Chicago: University of Chicago Press.
Latour, B. (1983) 'Give me a laboratory and I will raise the world'. In K. Knorr-Cetina & M. Mulkay (eds), *Science Observed: Perspectives on Social Studies of Science*. Berkeley, CA: Sage.
Latour, B. (1986) 'The powers of association'. In J. Law (ed.), *Power, Action and Belief: A New Sociology of Knowledge?* Sociological Review Monographs No. 32. London: Routledge.
Latour, B. (1987) *Science in Action: How to Follow Scientists and Engineers Through Society*. Cambridge, MA: Harvard University Press.
Latour, B. (1988) *The Pasteurization of France*. Cambridge, MA: Harvard University Press.
Latour, B. (1991) 'Technology is society made durable'. In J. Law (ed.), *A Sociology of Monsters: Essays on Power Technology and Domination*. Sociological Review Monographs. London: Routledge.
Latour, B. (1993) *We Have Never Been Modern*. Cambridge, MA: Harvard University Press.
Latour, B. (1996) *Aramis, or The Love of Technology*. Cambridge, MA: Harvard University Press.

Latour, B. (1999) *Pandora's Hope; Essays on the Reality of Science Studies*. Cambridge MA: Harvard.
Latour, B. & Woolgar, S. (1979) *Laboratory Life: The Social Construction of Scientific Facts*. Beverly Hills: Sage.
Lau, C. (1992) 'Social conflicts about the definition of risks: The role of science'. In N. Stehr et al. (eds), *The Social Construction of Technological Systems*. Cambridge, MA: MIT Press.
Law, J. (1994) *Organizing Modernity*. Oxford: Blackwell.
Luhmann, N. (1989) *Ecological Communication*. Chicago: University of Chicago Press.
Luhmann, N. (1993) *Risk*. Berlin: de Gruyter.
Lynch, M. (1993) *Scientific Practice and Ordinary Action: Ethnomethodology and Social Studies of Science*. Cambridge: Cambridge University Press.
Meadows, D.H., Meadows, D.L. & Randers, J. (1971) *The Limits to Growth. Report of the Club of Rome*. London: Earth Island.
Mol, A. & Law, J. (1994) 'Regions, networks and fluids: anaemia and social topology'. *Social Studies of Science*, 24, 641–71.
Munro, R. (1995) 'Governing the new province of quality: autonomy, accounting and the dissemination of accountability'. In A. Wilikinson & H. Willmott (eds), *Making Quality Critical*. London: Routledge.
Munro, R. (1996) 'Alignment and identity work: the study of accountants and accountability. In R. Munro & I. Mouritsen (eds), *Accountability. Power, Ethos and the Technologies of Managing*, pp. 1–19. Oxford: International Thomson Business Press.
Nelson, E. (1993) 'Ecolabelling'. *Environmental Liability*, 1(1), 16–19.
Newton, T. (1996) 'Agency and discourse: recruiting consultants in a life insurance company'. *Sociology*, 30(4), 717–39.
Newton, T. & Harte, G. (1997) 'Green business: technicist kitsch?' *Journal of Management Studies*, 34(1), 75–98.
Nietzel, H. (1996) 'Principles of product-related Life Cycle Assessment'. *International Journal of Life Cycle Assessment*, 1(1), 49–54.
Organization of Economic Co-operation and Development (OECD) (1995) An overview of the life cycle approach to product/process environmental analysis and management. Paris.
Porter, T. (1995) *Trust in Numbers: The Pursuit of Objectivity in Science and Public Life*. Princeton, NJ: Princeton University Press.
SETAC (1991) 'A technical framework for Life-Cycle Assessment'. Workshop report from the Smugglers Notch, Vermont, USA. Workshop held 18–23 August 1990.
SETAC (1992) Workshop report from the SETAC-Europe workshop held Leiden, Netherlands, December, 1991.
SETAC (1993a) *'Guidelines for Life-Cycle Assessment: A Code of Practice.'* SETAC, 1993.
SETAC (1993b) 'Conceptual framework for impact assessment'. Workshop report from the Sandestin, Florida, USA. Workshop held 4–9 October 1992.
Shackley, S. & Wynne, B. (1995) 'Global climate change: the mutual construction of an emergent science–policy domain'. *Science and Public Policy*, 22(4), 218–30.
Shapin, S. & Schaffer, S. (1985) *Leviathan and the Air Pump: Hobbes, Boyle and the Experimental Life*. Princeton: Princeton University Press.

Simmons, P. (1996) 'Eco-labelling and the construction of environmental risks'. In J. Holmwood, H. Radner, G. Schultze & P. Sulkunen (eds), *Constructing the New Consumer Society*. London: Macmillan.
Simmons, P. & Wynne, B. (1992) 'Responsible care, trust, credibility and environmental management'. In K. Fischer & J. Schot (eds), *Environmental Strategies for Industry*. Washington, DC: Island Press.
Star, S.L. & Griesemer, J.R. (1989) 'Institutional ecology, "Translations" and boundary objects: amateurs and professionals in Berkeley's Museum of Vertebrate Zoology, 1907–39'. *Social Studies of Science*, 19, 387–420, Sage Publications.
Suchman, L. (1994) 'Working relations of technology production and use'. *Computer Supported Cooperative Work (CSCW)*, 2, 21–39.
Sustainability, Spold & Business in the Environment (1993) *'The LCA Sourcebook: A European Business Guide to Life-Cycle Assessment.'* SustainAbility. 1993.
UKEB (1993) *Criteria for Hairspray Ecolabels: A Proposal by the UK Ecolabelling Board*. London: UK Ecolabelling Board.
United States Environmental Protection Agency (1993) *Life Cycle Design Guidance Manual*, EPA/600/R-92/226.
Van Den Belt, H. & Rip, A. (1997) 'The Nelson–Winter–Dosi model and synthetic dye chemistry'. In W. Bijker, T. Hughes & T. Pinch, *The Social Construction of Technological Systems*. Cambridge, MA: MIT Press.
Waterton, C. & Wynne, B. (1996) 'Building the European Union: Science and the cultural dimensions of environmental policy'. *Journal of European Public Policy*, 3(3), 421–40.
Wheeler, D. (1993a) 'Eco-labels or eco-alibis?' *Chemistry and Industry*, p. 260, 5 April 1993.
Wheeler, D. (1993b) 'What future for product lifecycle assessment?'. *Integrated Environmental Management*, 20, 15–19.
Wheeler, D. (1993c) 'Why ecological policy must include human and animal welfare'. *Business Strategy and the Environment*, pp. 36–8.
Woolgar, S. (1981) 'Interests and explanation in the social studies of science. *Social Studies of Science*, 11, 365–94.
Woolgar, S. (1991) 'Configuring the user: the case of usability trials'. In J. Law (ed.), *A Sociology of Monsters: Essays on Power, Technology and Domination*. London. Routledge.
Woolgar, S. (1996a) 'Technologies as cultural artefacts'. In W. Dutton (ed.), *Information and Communication Technologies – Visions and Realities*. Oxford: Oxford University Press.
Wynne, B. (1992a) 'Uncertainty and environmental learning: reconcieving science and policy in the preventative paradigm'. *Global Environmental Change*, 2(2), 111–27.
Wynne, B. (1992b) 'Misunderstood misunderstandings: social identities and public uptake of science'. *Public Understanding of Science*, 1(3), 281–304.
Wynne, B. (1992c) 'Risk and social learning: from reification to engagement.' In S. Krimsky & D. Golding (eds), *Social Theories of Risk*. Westport, CT: Praeger.

8

Perceived Justice and the Economic Valuation of the Environment: A Role for Fair Decision-Making Procedures

Bradley S. Jorgensen

Summary

Competing claims of fairness are often apparent in environmental planning decisions. With increasing awareness of pollution among various stakeholders, issues arise over appropriate and acceptable levels of service given finite public resources. The contingent valuation method has enabled economists to place environmental planning within a benefit–cost framework. However, justice issues have not been widely discussed in the CV literature despite evidence to suggest that survey participants scrutinize the fairness of the method, and its requirement that value be expressed in terms of WTP in particular. This chapter reviews some of the literature on fairness and suggests that a more deliberative decision-making framework might be required to account for the ethical beliefs about the environment, public institutions and the public interest. In providing citizens a greater opportunity to voice their views on environmental policies than the standard CV methodology allows, deliberative procedures and their outcomes are likely to be perceived to be more fair.

Introduction

National commitments to a cleaner, safer environment are subject to competing claims upon the public purse. Given finite public resources, difficult questions are inevitably raised. How clean should we make the air? Should water quality in an urban river be as clean as a pristine lake? How much damage to the environment should we tolerate from vehicle use?

Economists tackle these types of questions by balancing the monetary costs of a policy with its monetary benefits. When a policy concerns the quality or availability of a market commodity, the costs and benefits can be gauged from observing consumer responses to changes in prices. But, where the policy implicates environmental quality or some other public good, changes in prices or behaviors are not available since no markets exist. Rather, individuals' economic preferences for environmental improvements must be inferred from what people say they would be willing to pay if a market for trading environmental improvements actually existed. Instead of observing actual market preferences in the form of exchange values, people's intended purchases are measured by asking questions about each individual's willingness to pay (WTP) to obtain a cleaner environment. The use of surveys to elicit people's hypothetical monetary preferences for environmental benefits is called 'contingent valuation' (CV).

The allocation of scarce natural resources among competing demands in a society is an attempt to reconcile sometimes incommensurate values (Fiske and Tetlock, 1997; Vatn and Bromley, 1994). Trade-offs between alternative allocations can become difficult when social discourse comprises a pluralism of ethical positions having varying degrees of overlap and articulation. These instances can be viewed as conflicts involving different conceptions of justice. Central to notions of justice is a 'process for allocating resources according to a principle of deservingness' (Schwartz, 1975, p. 111). Difficulties arise, however, when principled statements about environment–social relations are so transcendent and abstract that they fail to find expression within the tightly circumscribed language of public policy and economics.

Types of justice

The concept of justice has been described as distributive, procedural, and interactional. Distributive justice focuses on perceptions of the outcomes that are derived from decision-making processes (Deutsch, 1975) and may be based on principles of equality (Sampson, 1975), equity (Walster, Berscheid and Walster, 1973), or need (Schwartz, 1975). Procedural justice concerns perceptions of the fairness of the decision-making process itself (Lind and Tyler, 1988; Thibaut and Walker, 1975). The concept of interactional justice involves the manner in which authorities conduct their relationships with other parties (Bies and Moag, 1986). Although it is useful to think of the three types of justice as separate concepts, they share a degree of overlap in practice (Folger, 1996).

Fairness Heuristic Theory (Lind, Kulik, Ambrose and De Vera Park, 1993; Van den Bos, Lind, Vermunt and Wilke, 1997) proposes that perceptions of procedural justice influence how the distribution of outcomes is evaluated. This *fair process effect* has been observed when participants have no information upon which to compare their own outcomes with those received by others. When social comparisons are possible, people rely less on procedural information (Van den Bos et al., 1997). Real world decisions are likely to advantage procedural information because unambiguous outcome information is often not readily available. Further, procedural information can provide greater insight about the trustworthiness of an authority than outcome information and tends to be accessible prior to information about decision outcomes.

Justice research has tended to show that fair decision procedures are those that enable stakeholders to express their viewpoints. Cvetkovich and Earle (1994) pointed out that giving participants a voice in decisions is most likely to result in fairness when they have played a role in defining the structure of the decision-making framework. This 'transformational voice' enhances the likelihood that the decision process will be consistent with participants' salient values. Given heterogeneity among the value positions held by different social groups, participatory decision-making is best viewed as a negotiation process (Cvetkovich and Earle, 1994; Syme and Eaton, 1989).

Values and contingent valuation

In their development of contingent valuation, neoclassical economists have sought to extrapolate the theory of private goods to the case of public goods and the environment (Green and Tunstall, 1991). In general, people may tend to believe that utilitarian market relations are an appropriate context in which to resolve the allocation of ordinary consumer goods, although research on the allocation of water has shown that people tend to think of natural resources as either private or public goods (Syme and Nancarrow, 2000). This disjunction between the perception of the good and the valuation context can contribute to problems of constitutive incommensurability and rejection of the trade-off decision. According to Fiske and Tetlock (1997) constitutive incommensurability refers to the moral limits to trade-offs when comparison among basic values threatens self-image and social identity. The authors refer to these trade-offs as 'taboo trade-offs':

> By a taboo trade-off, we mean any explicit mental comparison or social transaction that violates deeply-held normative intuitions

> about the integrity, even sanctity, of certain forms of relationship and of the moral–political values that derive from those relationships. (Fiske and Tetlock, 1997, p. 256)

A trade-off is likely to prove difficult or impossible when it calls for an evaluation of an environmental object in a manner that transgresses and challenges the social relationships that typically give it meaning. For example, people might adversely respond to the application of market prices to objects governed by communal social relations.

In contingent valuation there is a growing body of evidence to suggest that preferences for the environment do not represent trade-offs between income and environmental improvements. Rather, it has been proposed that preferences can be lexicographic for individuals who hold strong ecocentric beliefs (Edwards, 1986; Spash, 1997; Spash and Hanley, 1995). For example, Spash (1997) has argued that individuals who assign rights to the environment are likely to either reject the WTP question in CV or offer a zero bid. These responses would incorrectly be interpreted as participants placing no value on the environmental improvement. Alternatively, individuals may offer relatively large amounts of money that reflects their strength of feeling toward environmental protection rather than a trade-off between different states of the world (Ritov and Kahneman, 1997).

There is reason to believe that lexicographic preferences can also be apparent in CV independent of ecocentric belief systems (Jorgensen, 2000). After all, environmental improvements can implicate civil rights as well as ecocentric ones. People may scrutinize a decision process that (1) privileges those who can afford to pay, (2) fails to address perceived inefficiencies in relevant institutions, (3) restricts opportunities for meaningful participation in decision-making, and (4) rules out opportunities to discuss alternative funding arrangements (for example, existing public spending priorities, economic mechanisms that target particular social groups and not others, and so on). Note that any one of these factors can lead to discontinuous preferences without presupposing any strong environmental sentiment. Rather, absolute priority can be given to an attribute of the economic decision process (that is, its fairness) independent of an evaluation of the merits of the environmental policy. In this event, WTP reflects the value of the decision process, and principally the act of paying, rather than the value of the environmental improvement.

An argument can be made that the policy should include its funding arrangements, and this has been proffered to counter so-called payment vehicle bias (Mitchell and Carson, 1989). However, paying an additional sum of money cannot be considered part of a policy and also serve as the indicator of the value of that policy. If WTP doesn't support an assumption of indifference between choice alternatives then efficiency claims are unwarranted (Edwards, 1986).

Given that fairness concerns are often apparent in environmental planning decisions (Cvetkovich and Earle, 1994; Ebreo, Linn and Vining, 1996; Montada and Kals, 1995; Syme and Fenton, 1993; Syme and Nancarrow, 1996, 1997, 2000), it is surprising that justice issues are not routinely discussed in the CV literature. In fact, most published applications of the CV method give the impression that the greater bulk of participants are either willing to pay for valued environmental improvements or unwilling to spend money on policies that fail to deliver benefits consistent with their narrowly proscribed self-interests. In light of this, one might question the extent to which justice issues are prevalent in the contingent valuation of environmental policies.

Are fairness issues apparent in contingent valuation?

While individuals may hold attitudes toward particular environmental improvements, they may also have strong beliefs about the effectiveness of certain institutions, the causes of pollution, and/or ways of allocating the costs of pollution abatement. Put plainly, some individuals reject CV surveys by refusing to pay even though they value the environmental improvement. These 'protest responses' are troublesome in CV because they do not reflect individuals' values for the public good change presented in the survey (Lindsey, 1994). For this reason it is routine for a percentage of responses to be excluded from the data set on the basis that they are not consistent with a neoclassical interpretation of preference. At the root of this inconsistency are individuals' concerns about who should pay for environmental improvements and the fairness of the decision-making context presented by the contingent valuation survey (Jorgensen and Syme, 1995).

This claim was put to an empirical test in a study investigating WTP for stormwater pollution abatement (Jorgensen, Syme, Bishop and Nancarrow, 1999; Nancarrow, Jorgensen and Syme, 1995). The study involved face-to-face interviews with 1193 householders from Australian cities. Randomly selected households were stratified on three levels of income and two levels of proximity to a major water body. Photographs, maps and information cards were used to describe stormwater pollution and strategies that could be undertaken to alleviate it. Finally, respondents received either an open-ended (with benchmarked payment cards) WTP question format or the dichotomous choice version.

Table 8.1 Frequency of payment beliefs

Payment beliefs	% of N			χ^2
	Total ($N = 610$)	**DC ($N = 332$)**	**OE ($N = 278$)**	
1. I pay enough already	50.50	57.50	42.10	14.43***
2. I can't afford to pay at the moment	34.60	38.60	27.70	7.99***
3. Should use existing money	26.70	29.50	23.40	2.91
4. Those who pollute should pay	13.30	13.90	12.20	0.35
5. There is not enough information	12.00	4.50	19.10	32.32***
6. Unfair to ask me to pay anything	10.70	15.40	5.00	16.94***
7. It's my right to expect clean water	4.60	5.40	3.60	1.15
8. Don't want to place a value on water	3.80	3.90	3.60	0.04
9. Those who use the water should pay	3.00	3.30	2.50	0.33
10. It's not my problem	2.00	3.00	0.70	4.12*
11. Not a good way to deal with problem	1.80	1.20	2.50	1.47
12. That's all it's worth to me	0.03	0.30	0.40	0.02

* $p < .05$
*** $p < .001$

Protest responses were measured by asking respondents, who did not want to pay anything, to explain their responses. These rationales were post-coded into categories on a checklist by trained interviewers. Participants were free to cite as many reasons for their unwillingness to pay as they wanted to.

The frequencies of the payment beliefs are shown in Table 8.1. The most cited reason for not paying was the belief that one already paid enough for stormwater pollution abatement without the need for additional household payments. Chi-square tests indicated that some of the most oft-cited payment beliefs varied with the way the WTP question was asked. A significantly greater proportion of participants who responded to the dichotomous choice question format believed that (a) they already paid enough, (b) they could not afford to pay more, and

(c) it was unfair to have to pay anything. In short, the DC method was perceived as less fair than the open-ended version.

The study produced a number of additional results that are worth restating here. First, the results showed that the reasons individuals gave for refusing to pay were not independent beliefs, but clustered around two dimensions: procedural fairness concerns about one's own payment and distribution issues implicating the payment of others. Payment beliefs that reflected low procedural fairness were Beliefs 1, 2, 6 and 11 in Table 8.1. The beliefs associated with low distributive fairness were Beliefs 3, 4, 7, 8 and 9. Further analysis indicated that the low procedural fairness was significantly associated with the dichotomous choice question format, residing further away from a major waterbody, and lower household incomes. Low distributive fairness was also associated with the dichotomous choice format and lower household incomes, but not to participants' proximity to water.

The results of the study support the contention that participants evaluate the fairness of paying for environmental improvements. Notwithstanding, the methodology was limited to those participants who were unwilling to pay. Participants who were willing to pay were not taken into account despite the possibility that they too considered the fairness of raising public revenue through additional household payments. Similarly, the study stops short in providing a basis for understanding the relationship between attitudes towards paying and WTP.

Do fairness evaluations predict WTP?

A second study examined the relationship between fairness and dichotomous choice WTP for stormwater pollution abatement in four state capital cities in Australia. Each city sample was presented with stormwater pollution abatement scenarios that differed in terms of the payment vehicle, the intervention, the payment regime, and the institution assigned to manage the abatement strategy (see Table 8.2). Furthermore, fairness was measured with six attitude items which were rated on 5-point Likert scales by all respondents irrespective of whether or not they were prepared to pay for pollution abatement (see Table 8.3).

This study demonstrated that individuals' WTP for stormwater pollution abatement was influenced by their attitude towards the fairness of paying, and that this relationship was unaffected by variations in the payment vehicle and other factors included in Table 8.2 (Jorgensen and Syme, 2000). Moreover, the results indicated that, while individuals take into account the amount of money they are required to pay to

Table 8.2 Differences among the CV surveys

Characteristic	Sample			
	Perth	**Melbourne**	**Sydney**	**Brisbane**
Payment vehicle	Trust fund	Local government drainage levy	Membership fee and/or local government levy	Trust fund
Payment regime	Each year	Each year for 4 years	Each year	Each year for 4 years
Intervention	Sediment removal, revegetation, street sweeping, drain eduction	A continuous deflection system unit	Sediment removal, litter removal, litter baskets, street sweeping	Artificial wetland
Implementing institution	2 local councils, state water agency,quasi-government agency	2 local councils	Local council, and/or Centennial Park and Moore Park Trust	City council

Table 8.3 Description of attitude towards paying items

Item label	**Item description**
Protest 1	It is unfair to ask me to pay more money for stormwater pollution controls.
Protest 2	The government should use existing revenue to pay for storm water pollution controls.
Protest 3	We would be able to afford better protection of [receiving waters] already if the government did not waste so much money.
Protest 4	It is my right to have cleaner stormwater and not something I should have to pay extra for.
Protest 5	If the money was collected, I don't really believe that it would be spent on making the stormwater cleaner.
Protest 6	I pay enough already in government and council charges for cleaner stormwater.

obtain the environmental improvement, they also consider the degree to which the act of paying is fair. These two evaluations can be reasonably independent of one another and of equal importance in the decision process.

Taken together, the two stormwater studies support the view that justice concerns are apparent in some individuals' decisions about paying for environmental goods and that they have a robust influence on WTP responses. However, these studies only dealt with one type of environmental good, and one that most individuals would not be expected to have a great deal of familiarity with prior to learning about it in the CV survey. It is plausible to suggest that fairness issues might simply arise when individuals lack the means to construct meaningful responses to WTP questions. In this respect, negative attitudes towards paying may be constrained by the characteristics of particular public goods.

Do attitudes towards paying depend on the type of environmental good?

In an effort to identify the extent to which attitudes towards paying were unique to certain types of environmental issues (for example, stormwater pollution abatement), a study was carried out in Northern Wisconsin that dealt with four different issues (Jorgensen, Wilson and Heberlein, 2001). These were: Indian spearfishing, biodiversity, water quality and wolf reintroduction. The scope of the issues also varied in this study. For example, participants were asked to consider separately an increase in 300 and 800 wolves, and protection of spearfishing rights on a small number of lakes and on a much larger number. These variations in the scope and type of environmental proposals enabled an examination of the generality of fairness beliefs across environmental goods and levels of environmental change.

A sample of 617 participants were interviewed over the telephone about their WTP for each of the goods and levels of scope. Their beliefs about paying were also measured using belief statements and Likert scales. These beliefs were then analyzed using confirmatory factor analysis to ascertain how much of their covariation could be attributed to attitudes towards paying for each specific good, and how much was associated with a general attitude towards paying (see Figure 8.1). While some respondents might believe that paying for water quality was fair, but not wolf reintroduction, others might feel that paying additional amounts of money for any environmental good is unfair. The former situation attributes fairness as an issue pertaining to the specific good in question, whereas the latter suggests that fairness concerns might revolve around the indicator of value in CV, the intention to pay. To the extent that various beliefs about paying for spearfishing, wolves, water quality and biodiversity reflect the same general attitude towards paying,

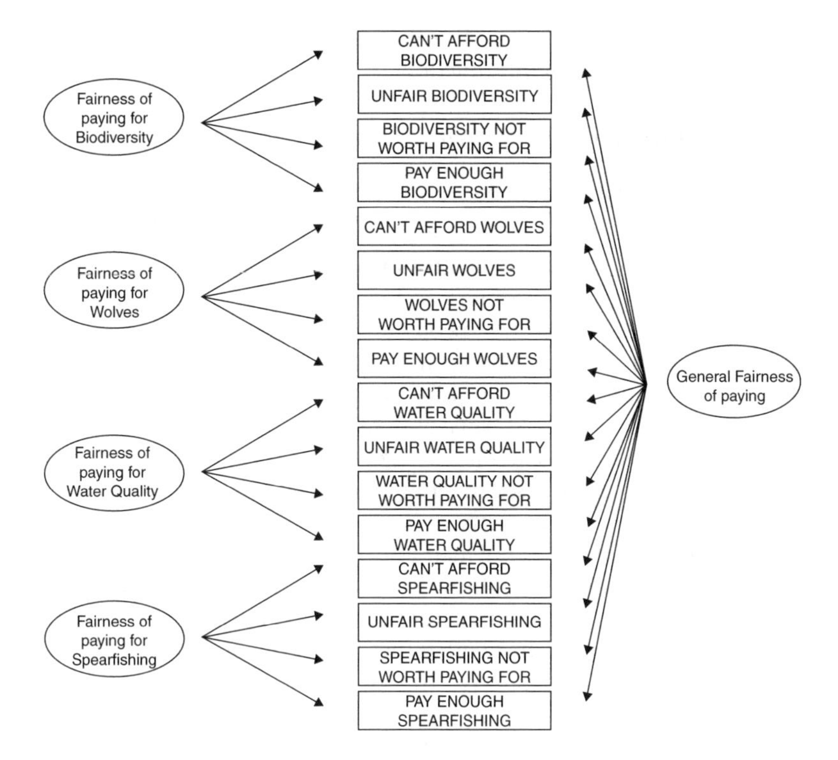

Figure 8.1 Confirmatory factor analysis model applied to each level of scope

then the type of environmental good is relatively less important in the way individuals think about the fairness issues in the act of paying.

The results showed that there was a significant degree of common variation across beliefs about paying that did not depend upon the environmental issue in question. This was true for both levels of scope. However, not all of the covariation among payment beliefs was associated with the general attitude towards paying. Rather, a significant amount of covariation also depended upon the specific good being valued. Beliefs about paying for spearfishing and wolves were associated with both the specific and general attitudes towards paying to about equal degrees. However, payment beliefs for biodiversity and water quality were much more a function of the general attitude than the specific one. This suggests that for these goods, beliefs about the fairness of paying was strongly consistent with an attitude towards paying for environmental goods in general. For wolves and spearfishing however, individuals thought about fairness in terms of their general attitude towards paying as well as concerns that were specific to Indian spearfishing and wolf reintroduction.

Improving the fairness of environmental valuation

Shifting the valuation context

Theorists have proposed (for example, Fiske and Tetlock, 1997; Thompson and Gonzalez, 1997) that trade-offs can be impossible for some individuals when framed in one context and yet possible when presented in a different context. For example, people seem quite open to the idea of participating in trade-off methodologies developed for urban planning and for setting public spending priorities (for example, Hoinville and Berthoud, 1970; Syme, Roberts and McLeod, 1990). These types of methodologies take into account budget constraints, but do not require that preferences be expressed in terms of an intention to make additional individual household payments.

Furthermore, Daniel Kahneman and his colleagues (Kahneman and Knetsch, 1992; Kahneman, Ritov, Jacowitz and Grant, 1993; Ritov and Kahneman, 1997) have pointed out that some individuals treat the CV context as a donation framework rather than a hypothetical market. To this extent individuals perceive the valuation context in a way that makes it more consistent with their understanding of the public good as a worthy cause in which social responsibilities entail.

While it is not possible to avoid the reality that public funds are limited, and that public revenue may need to increase to pay for new levels of service, contingent valuation alone offers little opportunity for individuals to participate in defining those environmental policy alternatives and the means of funding them. The difficulty with CV surveys is that they place marked constraints on the amount of information and deliberation that can be accomplished. The standard CV methodology requires that individuals, in isolation of peers and family, assign a dollar value to some environmental change following a limited passive information exchange. The inadequacies of the CV method in providing a context for meaningful participation in environmental decision making means that the WTP responses are likely to reflect a cursory and selective evaluation of the environmental issue at hand or scrutinize the CV survey itself as a method for ascertaining public values. When given the opportunity to choose between different ways of making public decisions, CV can rank at the bottom of the list of alternative methods (Oberholzer-Gee, Bohnet and Frey, 1997), regardless of participants' attitudes towards the method (Clark, Burgess and Harrison, 2000; Brouwer et al., 1999).

Deliberative methods of policy development and evaluation

There is an emerging viewpoint in CV literature that emphasizes deliberation and consensus-formation as an appropriate way of valuing public goods in a democracy (Sagoff, 1998). The idea of deliberative democracy – democracy in which agreement is reached by a process of reason, argument and consensus among free and equal citizens – is at the heart of many environmental decision-making methods (Royal Commission on Environmental Pollution, 1998). Researchers are beginning to think about restructuring the standard CV method according to deliberative techniques such as focus groups (Jorgensen, 1999; Lunt, 1999; Sagoff, 1998) and citizen juries (Brown, Peterson and Tonn, 1995; Sagoff, 1998; Ward, 1999). By adopting deliberative methods Sagoff has claimed that many criticisms currently leveled at CV could be overcome.

Sagoff (1998) points out that some problems with the validity of CV arise because the positions citizens put forward are principled views of the public interest and not the desires and wants of individual utility maximizers. That is, responses to WTP questions represent citizen opinions about what ought to be done as a society rather than the personal preferences of individual consumers. According to Sagoff, only through the adoption of deliberative decision-making techniques can arguments about the public interest be accounted for.

Public participation enables people who are potentially affected by environmental policy decisions an opportunity to express their views in their own terms (Crowfoot and Wondolleck, 1990; Cvetkovich and Earle, 1994; Renn, Webler and Wiedemann, 1995; Syme and Eaton, 1989). Moreover, in the best examples of public involvement, this is done by providing a context that brings competing interest groups into relation with each other and the agencies responsible for the development and implementation of environmental policy (Cvetkovich and Earle, 1994; Jones, 1994; Syme, 1992). In general terms, public involvement provides policy makers with a framework for dialogue with the community in a way that can foster constructive debate on planning issues. It is this forum in which principled views about the environment and the public interest can be voiced, debated, and deliberated upon. Individuals will be more likely to accept outcomes if they are given the opportunity to express their views in a forum in which they have played a role in constructing, and where decision makers offer clear explanations for decisions and treat people with respect. In providing citizens a greater opportunity to voice their views on environmental policies than the standard CV methodology allows, deliberative procedures and their outcomes have a greater chance of being perceived as fair.

Conclusion

Ethical values serve to undermine the neoclassical notion of preferences by usurping the assumption of indifference among choice alternatives. If participants reject the trade-off posed in the WTP question because of unfair decision procedures then the responses fail to reflect the economic value of either the good or the policy proposal. Rather, the responses reflect attitudes toward the act of paying, the indicator of economic value. These attitudes underpin beliefs about paying additional amounts of money for environmental goods, particularly those pertaining to the fairness of the act, trust in government, and perceptions of equity among different social groups (Jorgensen, Syme, Bishop and Nancarrow, 1999; Jorgensen and Syme, 2000; Jorgensen, Wilson and Heberlein, 2001).

The extent and potential for public participation via contingent valuation surveys is extremely limited in comparison to methods more firmly tied to the ideals of participatory democracy (Royal Commission on Environmental Pollution, 1998). Deliberative decision-making models and trade-off methodologies developed in the social sciences have the potential to inform policy questions regarding proposed changes in levels of service and their means of funding. Exploring the public's willingness to pay within a broader decision framework might enable environmental trade-offs to be examined and provide insight into public perceptions of acceptable environmental policy initiatives.

References

Bies, R.J. & Moag, J.S. (1986) 'Interactional justice: Communication criteria of fairness'. In L.L. Cummings & B.M. Staw (eds), *Research in Organisational Behaviour*. CT: JAI Press.

Brouwer, R., Powe, N., Kerry Turner, R., Bateman, I.J. & Langford, I.H. (1999) 'Public attitudes to contingent valuation and public consultation'. *Environmental Values*, 8, 325–47.

Brown, T.C., Peterson, G.L. & Tonn, B.E. (1995) 'The values jury to aid natural resource decisions'. *Land Economics*, 71, 250–60.

Clark, J., Burgess, J. & Harrison, C.M. (2000) '"I struggled with this money business": Respondents' perspectives on contingent valuation'. *Ecological Economics*, 33, 45–62.

Crowfoot, J.E. & Wondolleck, J.M. (1990) *Environmental Disputes: Community Involvement in Conflict Resolution*. Washington, DC: Island Press.

Cvetkovich, G. & Earle, T.C. (1994) 'The construction of justice: A case study of public participation in land management'. *Journal of Social Issues*, 50, 161–78.

Deutsch, M. (1975) 'Equity, equality and need: What determines which value will be used as the basis of distributive justice?' *Journal of Social Issues*, 31, 137–50.

Ebreo, A., Linn, N. & Vining, J. (1996) 'The impact of procedural justice on opinions of public policy: Solid waste management as an example'. *Journal of Applied Social Psychology*, 26, 1259–85.

Edwards, S.F. (1986) 'Ethical preferences and the assessment of existence values: Does the neoclassical model fit?'. *Northeastern Journal of Agriculture & Resource Economics*, 15, 145–50.

Fiske, A.P. & Tetlock, P.E. (1997) 'Taboo trade-offs: Reactions to transactions that transgress the spheres of justice'. *Political Psychology*, 18, 255–97.

Folger, R. (1996) 'Distributive and procedural justice: Multifaceted meanings and interrelations'. *Social Justice Research*, 9, 395–416.

Green, C.H. & Tunstall, S.M. (1991) 'Is the economic evaluation of environmental resources possible?' *Journal of Environmental Management*, 33, 123–41.

Hoinville, G. & Berthoud, R. (1970) Identifying preference values. Report on development work, social and community planning research, London.

Jones, A.P. (1994) 'Involving the public in water management'. *Water Environment and Technology*, July, 34–5.

Jorgensen, B.S. (1999) 'Focus groups in the contingent valuation process: A real contribution or a missed opportunity?' *Journal of Economic Psychology*, 20, 485–9.

Jorgensen, B.S. (2000) Perceptions of fairness and the explanation of perfect embedding in the contingent valuation method. Fairness & Cooperation: The International Association of Research in Economic Psychology (IAAREP)/ Society for the Advancement of Behavioural Economics (SABE) Conference Proceedings (pp. 211–15). Baden, Austria, 12–16 July.

Jorgensen, B.S. & Syme, G.J. (1995) 'Market models, protest bids, and outliers in contingent valuation'. *Journal of Water Resources Planning and Management*, 121, 400–1.

Jorgensen, B.S. & Syme, G.J. (2000) 'Protest responses and willingness to pay: Attitude toward paying for stormwater pollution abatement'. *Ecological Economics*, 33, 251–65.

Jorgensen, B.S., Syme, G.J., Bishop, B.J. & Nancarrow, B.E. (1999) 'Protest responses in contingent valuation'. *Environmental & Resource Economics*, 14, 131–50.

Jorgensen, B.S., Wilson, M.A. & Heberlein, T.A. (2001) 'Fairness in the contingent valuation of environmental public goods: Attitude toward paying for environmental improvements at two levels of scope'. *Ecological Economics*, 36, 133–48.

Kahneman, D. & Knetsch, J.L. (1992) 'Valuing public goods: The purchase of moral satisfaction'. *Journal of Environmental Economics and Management*, 22, 57–70.

Kahneman, D., Ritov, I., Jacowitz, K.E. & Grant, P. (1993) 'Stated willingness to pay for public goods: A psychological perspective'. *Psychological Science*, 4, 310–15.

Lind, E.A., Kulik, C.T., Ambrose, M. & De Vera Park, M.V. (1993) 'Individual and corporate dispute resolution: Using procedural fairness as a decision heuristic'. *Administrative Science Quarterly*, 38, 224–51.

Lind, E.A. & Tyler, T.R. (1988) *The Social Psychology of Procedural Justice*. NY: Plenum Press.

Lindsey, G. (1994) 'Market models, protest bids, and outliers in contingent valuation'. *Journal of Water Resources, Planning and Management*, 120, 121–9.

Lunt, P. (1999) 'Beyond measurement issues in the focus group method'. *Journal of Economic Psychology*, 20, 491–4.

Mitchell, R.C. & Carson, R.T. (1989) *Using surveys to value public goods: The contingent valuation method*. Washington, DC: Resources for the Future.

Montada, L. & Kals, E. (1995) 'Perceived justice of ecological policy and proenvironmental commitments'. *Social Justice Research*, 8, 305–27.
Nancarrow, B.E., Jorgensen, B.S. & Syme, G.J. (1995) Stormwater Management in Australia: Community Perceptions, Attitudes and Knowledge. Urban Water Research Association of Australia, No. 95.
Oberholzer-Gee, F., Bohnet, I. & Frey, B.S. (1997) 'Fairness and competence in democratic decisions'. *Public Choice*, 91, 89–105.
Renn, O., Webler, T. & Wiedemann, P. (1995) *Fairness and Competence in Citizen Participation: Evaluating Models of Environmental Discourse*. Dordrecht: Kluwer.
Ritov, I. & Kahneman, D. (1997) 'How people value the environment: Attitudes versus economic values'. In M.H. Bazerman, D.M. Messick, A.E. Tenbrunsel & K.A. Wade-Benzoni (eds), *Environment, Ethics, and Behaviour: The Psychology of Environmental Valuation and Degradation*. San Francisco: New Lexington Press.
Royal Commission on Environmental Pollution (1998) Twenty-first report: Setting environmental standards. Chairman, Sir John Houghton, Great Britain Royal Commission on Environmental Pollution.
Sagoff, M. (1998) 'Aggregation and deliberation in valuing environmental public goods: A look beyond contingent pricing'. *Ecological Economics*, 24, 213–30.
Sampson, E.E. (1975) 'On justice as equality'. *Journal of Social Issues*, 31, 45–64.
Schwartz, S.H. (1975) 'The justice of need and the activation of humanitarian norms'. *Journal of Social Issues*, 13, 111–36.
Spash, C.L. (1997) 'Ethics and environmental attitudes with implications for economic valuation'. *Journal of Environmental Management*, 50, 403–16.
Spash, C.L. & Hanley, N. (1995) 'Preferences, information and biodiversity preservation'. *Ecological Economics*, 12, 191–208.
Syme, G.J. (1992) 'When and where does participation count?' In M. Munro-Clark (ed.), *Citizen Participation in Government*. Sydney: Hale and Iremonger.
Syme, G.J. & Eaton, E. (1989) 'Public involvement as a negotiation process'. *Journal of Social Issues*, 45, 87–107.
Syme, G.J. & Fenton, D.M. (1993) 'Perceptions of equity and procedural preferences for water allocation decisions'. *Society and Natural Resources*, 6, 347–60.
Syme, G.J. & Jorgensen, B.S. (1994) *Assessing community values of capital works in the water industry: Contingent valuation and other techniques*. Division of Water Resources Consultancy Report No. 94/24.
Syme, G.J., Roberts, E. & McLeod, P.B. (1990) 'Combining willingness to pay and social indicator methodology in valuing public services: An example from agricultural protection'. *Journal of Economic Psychology*, 11, 365–81.
Syme, G.J. & Nancarrow, B.E. (1996) 'Planning attitudes, lay philosophies, and water allocation: A preliminary analysis and research agenda'. *Water Resources Research*, 32, 1843–50.
Syme, G.J. & Nancarrow, B.E. (1997) 'The determinants of perceptions of fairness in the allocation of water to multiple uses'. *Water Resources Research*, 33, 2143–52.
Syme, G.J. & Nancarrow, B.E. (2000) Fairness and its implementation in the allocation of water. Xth World Water Congress, 12–17 March 2000, Melbourne, Australia.
Thibaut, J. & Walker, L. (1975) *Procedural Justice: A Psychological Analysis*. NJ: Erlbaum.
Thompson, L.L. & Gonzalez, R. (1997) 'Environmental disputes: Competition for scarce resources and clashing of values'. In M.H. Bazerman, D.M. Messick,

A.E. Tenbrunsel & K.A. Wade-Benzoni (eds), *Environment, Ethics, and Behaviour: The Psychology of Environmental Valuation and Degradation*. San Francisco: New Lexington Press.
Van Den Bos, K., Lind, E.A., Vermunt, R. & Wilke, H. (1997) 'How do I judge my outcome when I do not know the outcome of others? The psychology of the fair process effect'. *Journal of Personality and Social Psychology*, 72, 1034–46.
Vatn, A. & Bromley, D.W. (1994) 'Choices without prices without apologies'. *Journal of Environmental Economics & Management*, 26, 129–48.
Walster, E., Berscheid, E. & Walster, G.W. (1973) 'New directions in equity research'. *Journal of Personality and Social Psychology*, 25, 151–76.
Ward, H. (1999) 'Citizen juries and valuing the environment: A proposal'. *Environmental Politics*, 8, 75–96.

Part V

Managing the Environment

9
Wise Use of Wetlands Tested in the Somerset Levels and Moors

A.R.D. Taylor

Introduction

The need to understand the environmental impact of my agronomic work on rain-fed rice in Kenya was the starting point for my work in wetlands. Later, while setting up the Uganda National Wetlands Programme, it was clear that wetland functions were poorly understood, although benefits were well known – for example, harvests of fish, papyrus, grasses, clean water supply. A strong interest in the benefits quickly sets an applied research agenda – for example, determining sustainable fish harvests through catch and return recruitment studies; randomized fixed point papyrus quadrats and cutting frequencies; water quality assessment in transects downstream from sewage discharges.

Around 1990, the Bureau for the Convention on Wetlands of International Importance especially as Waterfowl Habitat (Ramsar Convention) began collecting case studies of 'wise use' of wetlands. Wise use is simply a well-informed and sustainable use of resources. Among the key challenges to anyone interested in promoting 'wise use' is the need to be able to define the baseline hydrological and land-use regime that will allow the soils, and their dependent archaeological and biodiversity resources, to be conserved, that is, leaving future options open. Secondary challenges include the need to establish interdisciplinary working, and to establish political channels able to act upon the management information derived from research.

In many respects, people living in less developed countries practise wise use at a practical and unconscious level and conserve resources. Overwhelmingly, the money and resources devoted to wetland conservation comes from 'developed' countries, yet the 'developing' countries often practise more holistic management and use of wetlands than the former. In most western European countries, sectoral management for agriculture, forestry, fisheries, water supply and the conservation of landscape, biodiversity and archaeology do not provide a framework for successful wetland management. Management of wetlands is an interdisciplinary process cutting across all land use interests – if we can manage wetlands successfully then we have solved the problem of managing our entire environment.

In western Europe, similar wetland types can often be recognized across the region, resulting from a convergence of human lifestyle and common species that inhabit the wetland. Key factors which apply to European wetlands include the following:

- Dry land farming and forestry practices have often been imposed on wetland soils and hydrological conditions, rather than adapting these practices to undrained wetland.
- Most of the wetlands of western Europe are now no longer 'natural' with the possible exception of estuarine or shallow coastal waters.
- Most specialized marshes, meadows and fen habitats depend upon farming for their continued present-day economic or biodiversity value.
- Complex and expensive agri-environment measures dominate wetland systems, which are rarely research-driven.
- Community or 'peer management' of wetlands is rare.

Research-based management requirements

The resource: Somerset Levels and Moors wetland

The Somerset Levels and Moors is a coastal floodplain grazing marsh of about 64 000 ha, dissected by a largely man-made drainage system and up to the early twentieth century it was the largest intact wetland in south west Britain. The wetland to catchment ratio of 1 : 3.5 led to the formation inland of extensive sedge (containing for example, *Eriophorum angustifolium*, and *Cladium mariscus*) and moss peat deposits (*Sphagnum* sp.) now known to be traversed with many wooden trackways, the earliest dating from 4000 BC. The name of the county of Somerset originates from the ancient seasonal occupants of the wetland – *Sumorsaeta* or 'Land of the Summer People'. Intensive drainage and agricultural practices began to be adopted in the Somerset wetlands from 1960 onwards.

Waterfowl counts from 60 000 to 100 000 have been recorded regularly for the Levels and Moors, thus qualifying the site for designation under the Convention. The adjoining Bridgwater Bay National Nature

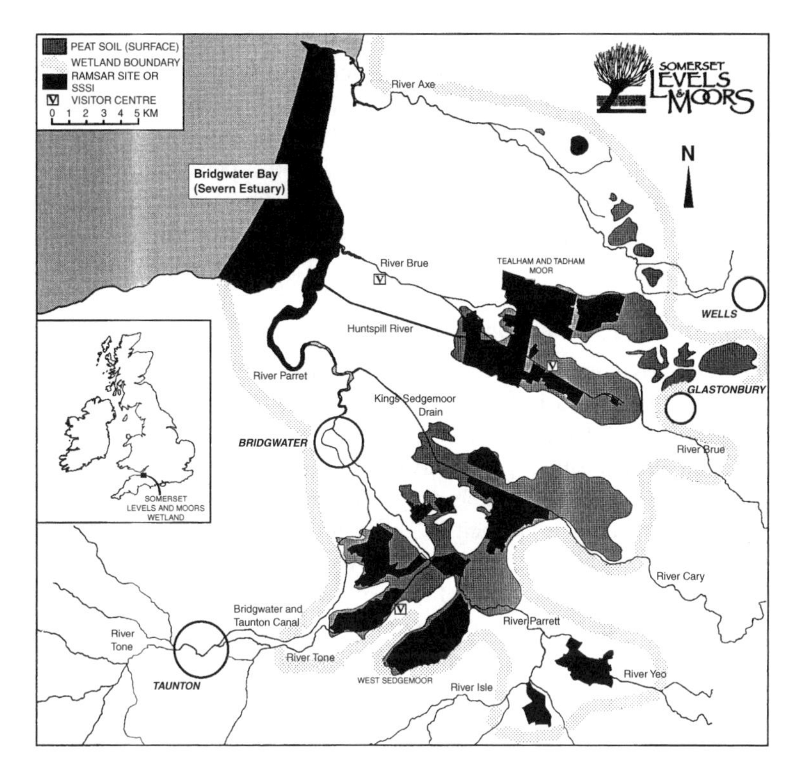

Figure 9.1 The wetland system of Somerset Levels and Moors

Reserve (2703 ha) is already designated, and most of the Levels and Moors Sites of Special Scientific Interest form a Ramsar site. Before the UK is able to satisfy its obligations under the Convention, good hydrological management demands the return of historic water levels of good quality over much of the area, especially the peat moors. Figure 9.1 outlines the main features of this wetland system.

Incentives available for wetland conservation and wise use

The most significant changes in the intensity of land-use have resulted from overproduction arising from the EC Common Agricultural Policy (partly reformed in 1992). In the UK, pressure from organizations concerned with monitoring the decline in wetland bird numbers, coincided with early EC wide initiatives to provide greater environmental protection to wetland bird habitat, primarily the 1979 Directive on the Conservation of Wild Birds. Later, Article 19 of the European Communities Structures

Regulation (797/85) was implemented in the UK by the Agriculture Act, 1986, which in the UK led to the identification of Environmentally Sensitive Areas, initially 10 in 1987 and now 22 which include the Somerset Levels and Moors wetland. Most of this wetland is in private ownership, therefore current incentives are aimed at persuading owners to restore environmentally sustainable farming practices and if possible, also to restore water levels to reverse the decline in biodiversity, particularly for breeding wetland birds (*Charadriiformes*).

Current land-use and key priorities

The cumulative impacts of fertilizers, pesticides, cultivation and reseeding with modern varieties has resulted in much increased livestock stocking rates. Dairy cattle, beef and sheep production now dominates the area, with a small proportion of arable crops. It has not been demonstrated that *any* arable crops are compatible with sustainable use of wetlands. The principal crops grown in Somerset wetlands include potatoes, wheat, high-yielding rye grasses and maize planted for silage production. Production of these crops depends upon use of fertilizers (200–350 kg N/ha), pump-drainage and ploughing which causes irreversible damage to peat-dominated soils. Peat shrinkage/oxidation is one indicator of damage and future lost opportunity, and it is thought that the rate of loss is up to 1 cm per year. Key recent work has shown that even with engineered raised water tables, peat may continue to oxidize (see, for example, Burton and Spoor, 1997; Gowing et al., 1997). Species loss is, however, dependent upon both drainage and fertilizer enrichment. Work done in the Somerset wetlands for English Nature, the Ministry of Agriculture Fisheries and Food (MAFF) and the former Department of the Environment (DoE) demonstrated that there is no safe level of fertilizer use that would conserve the biodiversity of wetland vegetation (Tallowin and Smith, 1994).

Key management priorities for the Somerset Levels and Moors include:

- raising of water levels to maintain peat hydration and reduce oxidation;
- allowing washland to regain its flood storage function;
- permitting more seasonal water flows across the floodplain to improve water quality;
- conservation of all traditional ‘unimproved’ wet grassland in the interests of biodiversity;
- elimination of fertilizer and pesticide use that impacts upon biodiversity and water quality;

- prevention of damage to buried archaeology;
- restoration of community oversight of wetland management coupled with dissemination of options for diversification and wise use.

Development of a research agenda

The Levels and Moors are not a historically unusual wetland system; many parts of the UK and western Europe share lowland floodplain systems. The UK and Ireland are, however, important in possessing much of the peat resources of western Europe, and the UK in particular has lost all but 4 per cent of its lowland mires (Cox, 1995). The research agenda reflects the pressure that such wetlands are under, and the international importance of their biodiversity, particularly for wetland birds. As pointed out earlier, wetlands require interdisciplinary management skills, and there is a high degree of hydrological connectivity between adjacent land uses. The above management priorities set the agenda for research.

The agenda therefore has three main components:

Environmentally sustainable water management

The overall water budget balance sheet for the Somerset Levels and Moors is unknown. In addition, the wetland is biphasic; in winter water may rise sufficiently to flow overland, but at most times of year, water is confined to the extensive ditch system, which drains into main rivers in winter, but is irrigated from the rivers in summer. Water tables in the rectangular fields vary from ponded, due to impeded drainage resulting from pan conditions, to sharp evapotranspirative drawdown in their centres, even while ditch levels may be high. Water management is thus guesswork and demands focused work to determine:

- minimum required water levels and flows and hydraulic conductivity;
- seasonal minima to ensure peat block rehydration and adequate water quality;
- determinants of water quality, especially in ditches.

Recent baseline research by Burton and Spoor (1997) and by Gowing et al. (1997) has shown that peat soils in the Brue Valley, Somerset have formed impermeable pans and cracks and have shrunk irreversibly. More worryingly, current raised water level areas funded by MAFF Environmentally Sensitive Area payments under Tier 3, may not be acting to conserve peat.

Water quality work has recently shown that most ditches are hypertrophic, yet clear water is the norm. More work is required to find out why.

These issues have been considered in the Water Level Management Strategy and Action Plan (Environment Agency, 1999) which aims to refine the practical actions that may be taken by all partners.

Conservation and enhancement of biodiversity

Biodiversity has been in sharp decline on the Somerset Levels and Moors, and while wintering wetland birds regularly exceed 60000, in 1997 breeding waders such as snipe, red shank, curlew and lapwing were 58 per cent lower than in 1977 (see Figure 9.2) except on small areas of ESA Tier 3 raised water level area. RSPB and others have shown that careful management of winter and early spring splash-flooded areas can promote successful breeding (see, for example, Chown, 1998), however, there are now concerns about losses in sward biodiversity in such conditions. These concerns have arisen due to the apparent effects of relatively uniform flooding and lack of flow in spring on Tier 3 raised water level areas. Key research is needed into:

- optimum water regime and habitat diversity capable of maintaining and enhancing breeding wader birds, soil invertebrates and meadow herbs and grasses;
- potential of damaged peat in lowered fields to support non-farmable species of high biodiversity interest.

Social and economic

The wetland remains largely in private ownership, except for over 50 per cent of West Sedgemoor (RSPB owned) and patches of National

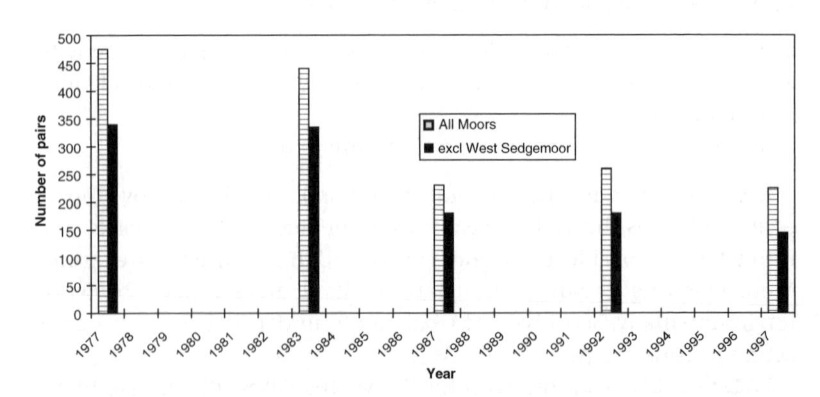

Figure 9.2 Total waders at 12 sites 1977–97 (RSPB data)

Nature Reserve, scattered across several moors. The special case of the Avalon Marshes, where the peat production zone is gradually reverting to deep water wetland, is not typical of the Somerset Levels and Moors, but does form an exciting case study of transition in land-use, where over 50 per cent is in public ownership.

Wise use depends upon effective public partnerships with private owners. In the ESA, which covers about 40 per cent of the wetlands, over 1000 voluntary agreements are in force (see Figures 9.3 and 9.4), guided by expert input from FRCA, English Nature and Environment Agency staff.

Yet because ESA agreements are not whole farm agreements, the coverage is only just over 55 per cent of the total ESA area of 29 000 ha. Outside the ESA there is very little incentive to promote wise use, and little understanding of the potential for land-use diversification. Economic valuation of wetlands is relatively new. Barbier et al. (1997) give valuation examples for wetland products and wetland functions, yet such valuation does not form part of 'income foregone' based MAFF valuations for determining ESA agreement payments. Recognition of the whole wetland, rather than its component SSSIs/SPA sites is being researched, and the special recognition of the wetland system as a 'Cultural Landscape' may provide its communities with better understanding of its

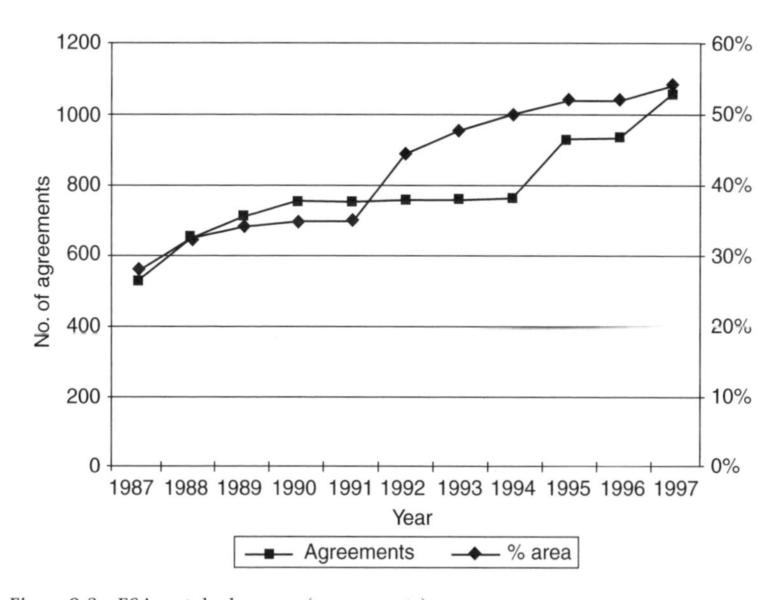

Figure 9.3 ESA uptake by year (agreements)

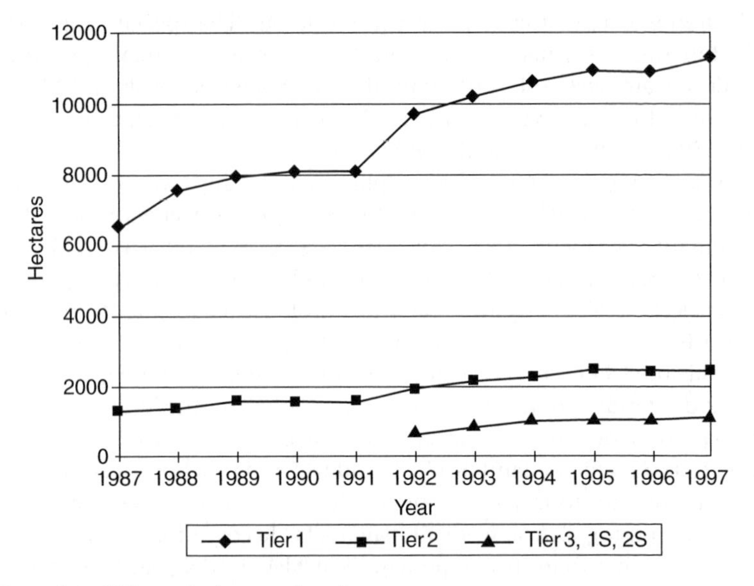

Figure 9.4 ESA uptake by year (area)

values and broad based economic incentives (Somerset County Council, 1999). Key research is needed to:

- establish tools for biodiversity valuation, aimed at focusing private land-use on achievable habitat or species targets;
- undertake flood plain service valuation, to determine whether a new floodplain management agreement is viable;
- identify wetland products which could diversify wetland land use (for example, willow coppice, fish, domestic waterfowl harvest).

Actions towards wise use – some examples

Key suggestions for promotion of wise use in Somerset include:

- promote awareness in economic terms, of the benefits to be obtained by diversification of wetland use (for example, the wetland's ability to support fish breeding, provide thatching, basket-making materials and other woody coppice crops and grazing resources that do not depend upon inorganic fertilizers or annual reseeding);
- maximize ecological tourism (or green tourism);

- reform or abolish Internal Drainage Boards, which have a duty to drain, but not to rehydrate wetland;
- review the long-term costs of management agreements for private land to determine whether it is better value to buy land to be transferred back to the community and managed by them, as demonstrated by the success of the Marais du Cotentin et du Bessin (see Lorfeuvre, 1993) in managing its wetland common land, set amongst private land holdings;
- transfer the involvement of local people from the state-funded land drainage system to positive management structures.

References and further reading

Barbier, E.B., Acreman, M. & Knowler, D. (1997) *Economic Valuation of Wetlands. A Guide for Policy Makers and Planners*. Gland, Switzerland: Ramsar Convention Bureau.

Bartlett, L. (1998) *An evaluation of water level management within the Somerset Levels and Moors and its effect upon agriculture, wildlife and its habitat.* BA thesis: Environmental Policy and Management 1998. London: Guildhall University.

Brunning, R. (1999) *Water Level Monitoring at Harter's Hill, Queen's Sedgemoor.* Somerset County Council, Taunton.

Burton, R. & Spoor, G. (1997) *Peat soil conservation in Somerset, part 2. Site investigations on bumpy fields on peat soils in the Brue Valley.* Silsoe College, Cranfield University. Report to English Nature, Roughmoor, Taunton.

Chown, D.J. (1998) *Somerset Levels and Moors breeding wader survey 1997.* RSPB, Dewlands Farm, Somerset.

Cox, M. (1995). 'Peat restoration and rehabilitation: ecological gain and archaeological loss?' In M. Cox, V. Straker & A.R.D. Taylor (eds), *Wetlands: Nature Conservation and Archaeology*. Proceedings of an International Conference held at the University of Bristol, 11–14 April 1994. London: HMSO.

English Nature (1999) *Conservation Management Plans for Selected Raised Water Level Areas*. Taunton: Roughmoor.

Environment Agency (1999) *Somerset Levels and Moors water level management strategy and proposed action plan consultation document.* Environment Agency, North Wessex Area, Bridgwater, Somerset (or from *www.somerset.gov.uk/levels*).

Gingell, C. et al. (1993) *Towards a sustainable future for the Levels and Moors.* Report for the 1993 North American/UK Countryside Exchange. Available from Somerset County Council, UK (or from *www.somerset.gov.uk/levels*).

Gowing, D., Gilbert, J. & Spoor, G. (1997) *Peat soil conservation in Somerset, part 3. Investigations into changes in ground surface levels, Tadham Moor experimental site, Somerset Levels and Moors*. Report to English Nature. Taunton: Roughmoor.

IGER, Somerset Wildlife Trust (1999) *Catcott Lows. Conservation of peat soils on the Somerset Levels and Moors. Part 5*. English Nature, Roughmoor, Taunton.

Lorfeuvre, F. (1993) 'Developing a wise use strategy for the Cotentin and Bessin Marshes'. In T.J. Davis (ed.), *Towards the Wise Use of Wetlands.* Ramsar Convention Bureau, Gland, Switzerland.

Mountford, J.O. & Barratt, D.R. (eds), with Manchester, S.J., Sparks, T.H., Treweek, J.R., Tallowin, J.R.B., Smith, R.E.N., Gilbert, J.C. & Gowing, D.R.B. (1999). *Wetland Restoration: techniques for an integrated approach. Phase IV: survey and experimentation.* Interim report from ITE to the Ministry of Agriculture.

Mountford, J.O. & Manchester, S.J. (eds), with Barratt, D.R., Dunbar, F.M., Garbutt, R.A., Green, I.A., Sparks, T.H., Treweek, J.R., Gilbert, J.C., Gowing, D.J.G., Lawson, C.S. & Spoor, G. (1998). *Assessment of the effects of managing water-levels to enhance ecological diversity.* Fourth ITE report to the Ministry of Agriculture.

Somerset County Council (1999) *Study of Special Recognition for the Somerset Levels and Moors.* Land Use Consultants, London, for Somerset County Council, Taunton, February 1999 (brief available from *www.somerset.gov.uk/levels).*

Swetnam, R.D., Mountford, J.O., Armstrong, A.C., Gowing, D.J.G., Brown, N.J., Manchester, S.J. & Treweek, J.R. (1998) 'Spatial relationships between site hydrology and the occurrence of grassland of conservation importance: a risk assessment with GIS'. *Journal of Environmental Management*, 54(3), 189–203.

Tallowin, J.R.B. & Smith, R.E.N. (1994) *The effects of inorganic fertilisers in flower-rich hay meadows on the Somerset Levels.* MAFF/NCC/DoE Tadham Project 8th report. English Nature Research Report, Peterborough, UK.

Treweek, J.R. (1993) *Wetland restoration: techniques for an integrated approach.* Phase II report. MAFF/NERC Contract TO2059 fl and MAFF/ADAS Contract BD0202. Institute of Terrestrial Ecology, Monks Wood, UK.

Treweek, J.R., Mountford, J.O., Manchester, S.J., Swetnam, R.D., Brown, N.J., Caldow, R.W.G. & Hodge, I.D. (1999) Research on the restoration of lowland ESAs. In *United Kingdom Floodplains. Proceeding of a symposium held at the Linnean Society*, 25–27 October 1995. London: Linnean Society, RSPB and NRA.

10

Environmental NGOs, Civil Society and Democratization in Eastern Europe

Caedmon Staddon

Summary

In this paper the author submits some common understandings about environmental non-governmental organizations (NGOs) in Eastern Europe to theoretical and empirical scrutiny and finds them wanting. The view that the growth of the environmental NGO sector in Eastern Europe is positively correlated with the development of 'civil society' and democratization has been constituted through the hegemonic representations of the West's own myth of becoming. It is thus doubly inappropriate, for it both distorts our understanding of post-communist realities and deludes us about ourselves. Through the presentation of some detailed case study material from ongoing Bulgarian field research between 1992 and 1997 he exposes some of these distortions and argues that environmental NGOs are more complex and therefore more worthy of study than the mainstream literature gives them credit for. In conclusion, he suggests a number of avenues for the development of interdisciplinary research in the role(s) of environmental NGOs in post-communist transition.

Introduction

Today I will speak about one part of a five-year research programme that I undertook while at the University of Kentucky in the US. Much of what I am going to present to you is also contained in a rather more detailed paper, co-authored with Barbara Cellarius of the University of Kentucky, that compares theorizations of civil society development in Eastern Europe with what is actually happening in Bulgaria where we have pursued independent research programmes. In keeping with the remit of the series I shall place this research within a broader context, mapping out a series of research questions about the roles of non-governmental organizations (NGOs) in environmental management in Eastern Europe and their interpretation in some quarters as indices of the health of something called 'civil society'. I shall dip into the details contained in the written paper as necessary. As you will see this presentation does tack back and forth between theory and empirics. I will begin by saying a little about how these research interests were developed.

My introduction to Eastern Europe and the scholarly study of transition came in 1991 when I began Ph.D. studies at the University of Kentucky. At that time I was a research assistant on a three-year research project funded by the John D. and Catherine T. MacArthur Foundation looking at regional restructuring in the coastal Bourgas region of eastern Bulgaria (research published in Paskaleva et al., 1998). That research programme was divided into four sub-units or working groups looking at *industrial restructuring, agricultural sector reform, tourism development* (which is an important part of the economy of the Black Sea coast) and *local environmental regulation*. In 1992 and 1993 I worked primarily with the first and last of these sub-groups developing an interest in the changing political relations between the local and the central states during the post-communist transition. Subsequent to that I developed a detailed case study of how emerging patterns of local/central relations in Bulgaria were manifested in a specific case of local environmental management; one that involved the resolution of a conflict over scarce drinking water supplies (Staddon, 1996). The completed dissertation focused on various sorts of institutions and processes that mediated between the central state (acting in the interests of the water scarce capital region) and the particular locality that was being tapped as a source of new supply. Particularly important in this case study were the NGOs and civil associations, some of which were legally registered and professionally recognized and some of which were more spontaneously generated and perhaps therefore somewhat more ephemeral. This aspect of the research raised for me the questions that I will explore today; questions about the growth and development of environmental NGOs in Eastern Europe and their relations, necessary and contingent, with the broader processes of post-communist transition.

This presentation is divided into four parts. I shall start with a brief sketch of why the development of the NGO sector is important for theory and policy in Eastern European environmental management. I will then introduce you to the social and economic realities of post-communist transition in Bulgaria, taking particular pains to go beyond mere economic statistics in an effort to communicate the local, even domestic, realities of transition. These are often overlooked in the scholarly literature, which in my view is often over-fixated on abstractions rather than real events affecting real people (Staddon, 1998). I will then turn to the roles played by the emergent environmental NGO sector in environmental assessment and clean-up since 1989. There I will compare the theoretical expectations outlined below with the realities as Barbara Cellarius and I have encountered them in our respective fieldwork programmes. I will conclude with some prospective thoughts about how these insights can be integrated into an interdisciplinary environmental research agenda.

Environmental degradation and the return of civil society in Eastern Europe

As Western scholars researching different areas of the transition in Central Eastern Europe, we have noted with some dismay the marked tendency to interpret the growing activities of non-governmental organizations (NGOs) in an uncritical light (cf. Cellarius and Staddon, 2002). Specifically, we are concerned about the tendency to assume a strong positive correlation between size of the NGO sector and the relative progress of democratization. This argument usually takes the form of the claim that NGOs are a sort of 'index' of the development of civil society which is in turn a critical element in the democratization process. For many Central and Eastern European states, there is the additional imputation that because environmental protest groups (often the precursors of today's environmental NGOs) played such a large role in bringing about the dramatic regime changes of 1989–90, these same organizations must therefore constitute the vanguard of democratic transition. We see this logical progression reflected fairly clearly in the literature. Indeed, the association between environmental NGOs, civil society and democratization in Bulgaria is so strong that Snavely and Desai (1995) have recently claimed that 'the rapid growth of environmental and nature conservancy groups is perhaps the most graphic illustration of the role of non-profit voluntary organizations in creating a civil society in Bulgaria'. Other proponents of this line of argument include Jancar-Webster (1993), Yancey and Seigel (1994) and the contributers to Diamond and Plattner (1992). Based on our extensive field work in the traditions of cultural anthropology and political geography, I would like to suggest this conceptual association, while fairly common, is highly tenuous and open to dispute.

Basically, there are two different sets of reasons why we doubt the veracity of this association between the growth of the environmental NGO sector and the establishment of a robust civil society in Eastern Europe. The first is *empirical* and is probably the most important; certainly it is the most immediate. The material presented here underlines the point that the NGO sector in Bulgaria is actually much more complex and varied than Western scholarship gives it credit for. Unfortunately there is no single identity position occupied by NGOs in the processes of post-communist change. A single NGO may play very different, and even contradictory, roles in the socio-cultural construction of environment issues, the political mediation of competing interests and the subsequent management of specific technical processes (cf. Staddon, 1996; Cellarius, 1998). These matters will be taken up shortly in the section treating the characteristics of the environmental NGO sector in Bulgaria. The second set of reasons why we think the model is suspect is *theoretical* and concern ongoing debates about the relationships between civil society, democratization, democratic institutions and citizenship. I am referring to the sorts of debates that were spurred by books like Jean Cohen and Andrew Arato's (1990) *Civil Society and Political Theory*, John Keane's (1988) *Democracy and Civil Society*, and other books in political science and political philosophy. Glossing over a tremendously complex literature I would suggest merely there is little agreement about the correct theorization of the civil society concept, of its relation to democratic decision making, and consequently little agreement about the adequacy of social phenomena such as NGOs as simple indicators of civil society.

Consider the theoretical relations between civil society and the state for a moment. Modern Western societies, societies that have reached the contemporary apotheosis of liberal capitalism, are by definition said to be market-driven with the institutions of the state existing primarily *as a brake* on the excesses of material privation generated by the free market system (the welfare function) and *as a guarantee* of the social acceptance and reproduction of that system (the social integration function). Eighteenth- and nineteenth-century political theorists such as Rousseau and Mill further argued that the state was an instrumental institutionalization of popular will, or the civil society (echoes of Hegel to be sure!). Paradoxically then, the modern liberal state therefore was held to be both subordinate to the realm of private non-market association, that is, *civil society*, and superordinate insofar as the latter was supplied with legal guarantees and protection by the former (Keane, 1988; Seligman, 1992; Bernhard, 1993). More recently some political theorists have suggested that civil society can best be reconceived as a 'setting of settings' within which other forms of social, political and economic activity take place. Walzer (1992) explicitly argues that what is at stake in the civil society debate is the recognition of the need to protect the spheres within which association can take place, rather than the associationalism *per se* – what Walzer refers to as 'critical associationalism'. The important point here is that we have moved some distance away from specific phenomena, such as environmental NGOs, and more towards a more fissiparous 'setting of settings'.

So the relations between civil society development and the rise of environmental NGOs in Eastern Europe may still be important, though they are unlikely to be simple. The task here is to thoroughly debunk any lingering tendencies towards equating the one with the other. For the 'NGO as Civil Society Model' model to be upheld, arguably four conditions have to hold empirically. *First* there must have been since 1989 a rise in the number of officially recognized NGOs as well as an increase in the diversity of their activities. *Second,* there must be evidence of a marked increase in the number of Bulgarians involved in these organizations which would suggest their increasing embeddedness in Bulgarian society. *Third,* there must be a framework of laws and institutions to protect and nurture the activities of this sector. *Fourth,* these organizations must practise some form of internal democracy that maximizes the involvement of all participants in decision making and implementation. This fourth point is the one that Barbara and I have had the most dialogue and debate about. For me it points in the direction of *discursive formations*; that is how these NGOs are conceiving environmental problems as problems capable of certain kinds of mediation. In my understanding of this fourth point there is a strong impetus to look at the social construction of environment and nature in NGO discourse, though I will stress that this line of thinking is at present incomplete. I invite your comments!

I suggest that there are at least two reasons why there is much to be learned from such a subject. The first is very pragmatic and policy orientated and the second is theoretical and exposes some important features of our Western understanding about post-communist transition. The pragmatic reason is that millions of dollars are being thrown at these organizations by international development agencies such as the World Bank, the PHARE programme, the EBRD and the whole cast and crew of the international development set. So we need to look very carefully at what exactly NGOs are expected to do by the donor bodies. While Creed and Wedel (1997) have recently suggested that Western expectations may be unrealistic given the path dependencies of post-communist societies, others such as Sampson (1996) and Cellarius (1998) have focused more on the ways in which NGOs have tended to constitute environmental 'problems' and 'projects'. Similarly, I really do think that at the end of the day, we want to be able to say something about our models of democratization and civil society development based on carefully circumscribed field research; research conducted over long periods in the countries of Eastern Europe, in their languages, and with due recognition of their specific conditions.

Bulgaria in transition: 1989–97

Before delving into the details of Bulgarian environmental NGOs, a quick overview of contemporary Bulgaria and especially its experience of the post-communist transition is in order. Bulgaria is one of the twelve Central and Eastern European countries that are currently undergoing a series of systemic transitions; economic, political, social and even cultural. It is located physically in the south eastern corner of the European continent and is often referred to by Bulgarians as an international 'cross-roads', linking Europe with Asia (Map 10.1). I think partly because of this peripheral geographic location, we in the West have tended to know much less about Bulgaria than we have about Central European countries such as the Czech Republic, Slovakia, Poland and Hungary (cf. Paskaleva et al., 1998).

According to the 1994 census, Bulgaria is a country of about nine million people of which about 85 per cent are ethnic Bulgarians, roughly 10 per cent are ethnic Turks whose forebears emigrated to the region while it was part of the Ottoman Empire, roughly 3 per cent are Roma peoples (the so-called 'gypsies'), while the remaining few per cent comprises a number of tiny minority groups such Tartars, Armenians and Circassians.

Physically it is a relatively small country, about half the land area of the United Kingdom. To put it in other terms, the run from my apartment in the capital Sofia to the Black Sea coast in my reliably unreliable Trabant 'Combi' (a two cylinder, two stroke horror!) takes about five hours; so you can pretty much cross the country in five or six hours. North to south it is a little bit harder because rather inconveniently the Balkan Range of mountains bisects the country on a west to east axis. Other mountain groups such as the Rhodope, Rila and Pirin are located in the south and south-west of the country.

Economically, the country has gone through very severe economic shocks since 1989, which are reflected in the statistics presented in

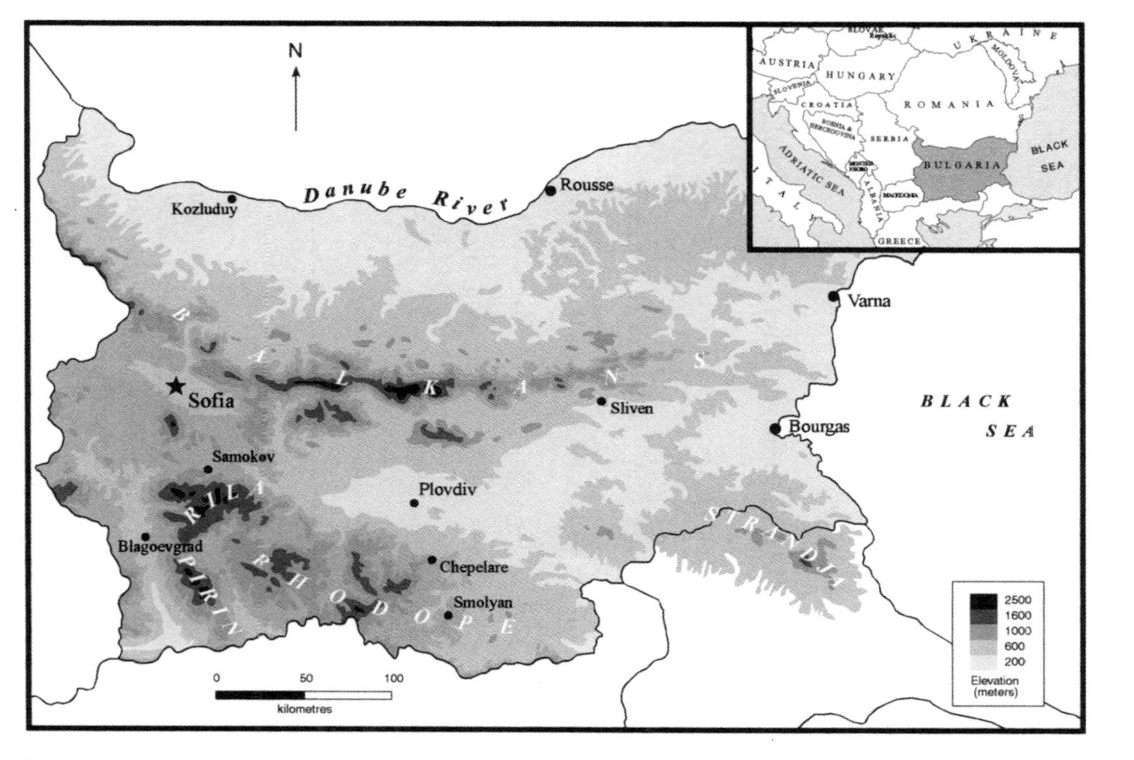

Map 10.1 Bulgaria in European context

Table 10.1 Bulgarian economic indicators, 1989–97

	1989	1990	1991	1992	1993	1994	1995	1996
GDP (billion LEVA)	35.6	45.4	131.0	195.0	299.0	543.0	867.7	1700.0
GDP (billion USD)	17.6	6.9	7.5	8.4	10.8	10.0	13.0	10.8
GDP per capita (USD)	1957.0	769.0	836.0	990.0	1280.0	1184.0	1546.9	1220.0
Private sector % of GDP	n.a.	9.1	11.8	15.3	19.4	28.0	–	–
Unemployment (1000s)	0.0	65.0	419.0	577.0	626.0	488.0	–	478.8
Unemployment rate (%)	0.0	1.6	10.8	15.5	16.4	12.8	10.8	12.5
Avg month wage (LEVA)	274.0	378.0	1012.0	2047.0	3145.0	4708.0	7597.0	12 290
Avg month wage (USD)	136.0	58.0	58.0	88.0	114.0	87.0	109.0	77.8
Annual change (%/yr) in:								
Real GDP	−1.9	−9.1	−11.7	−5.7	−2.4	1.4	2.1	−9.0
Gross indust'l production	−1.1	−16.8	−22.2	−15.9	−6.9	4.5	–	−6.0
Gross agri production	0.8	−6.0	0.0	−12.0	−18.2	0.8	–	–
Gross fixed investment	−0.5	−25.1	−15.6	−26.3	−29.7	n.a.	–	–
Consumer price inflation	10.0	72.5	338.9	79.6	64.0	121.9	132.9	410.8
Producer price inflation	n.a.	n.a.	284.0	24.9	15.3	91.5	–	257.1
Nominal average wage	8.7	38.0	167.7	102.3	53.6	51.4	–	161.0

Source: Wyzan, 1996; *Bulgarian Economic Review*, various dates 1997.

Table 10.1. If you look at GDP denominated in Bulgarian currency, things look rather rosy until you realize the obvious point, noted in the row third from the bottom, that Bulgaria has suffered from chronic high inflation since 1989. More reliably, trends in GDP measured in US dollars show that only by 1996 had GDP clawed its way back to about two-thirds of 1989 levels. Per capita GDP has also fallen, indicating a fall in the real productivity of the labour force, while unemployment has of course increased. (There are huge debates, by the way, about the reliability of such measures of unemployment.) The average monthly wage currently is US$78 per month, or something like £50. My colleagues at the Institute of Geography at the Bulgarian Academy of Science earn that princely sum for their monthly toils. Real GDP is down quite significantly. These and the other indicators in Table 10.1 speak abstractly for a general and marked economic downturn, but also I want to communicate to you the real impact this has had on everyday life for average Bulgarians. Times are very hard for people, and in order to get by a lot of Bulgarians have actually re-established connections with rural land, especially agricultural land, and there is now a huge flow of foodstuffs grown on tiny rural plots back into the urban areas, just to make up the bare domestic consumption fund. Indeed, processes of economic transition seem to be carving a wider and deeper gulf between the core regions of post-communist countries and their peripheries (Staddon, 1998).

Moving now from economic to environmental condition, virtually every part of the country suffers from some kind of environmental stress. Most of the major urban centres suffer from atmospheric, water and ground pollution caused by centrally planned infrastructural and industrial development. Nuclear contamination is endemic around the Kozludoy nuclear energy plant on the Danube River, and ferrous and non-ferrous metallurgical plant pollution problems exist around Sofia, Vratsa, Rousse, Bourgas, Kurdzhali and Plovdiv. There are serious and significant problems with the management of industrial, domestic and hazardous wastes, represented by the trash cans placed near most major urban centres, and virtually every region of the country suffers from some sort of water shortage. Since 1989 shortages of drinking water supplies in particular have generated their own environmental politics. Table 10.2 provides some more precise statistics for atmospheric pollution in the urban areas already mentioned. Pollution patterns are, I would suggest, fairly obvious. For example, the country's two lead–zinc smelters are obviously located at Kurdzhali and Assenovgrad, and the non-ferrous metallurgical plant at Pirdop shows up clearly as well. Also evident from these statistics is the fact that while overall levels of atmospheric pollution have declined for particulates attributed to lead and other atmospheric pollutants, the decline has not been as much as might have been anticipated, given the concomitant collapse in industrial production. Second, indicators for most of these substances are still above legally mandated maximum permitted concentrations (MPCs). Thus these reductions are hardly the product of a beneficial restructuring of industrial production, except in the one case of the Assenovgrad lead–zinc smelter

Table 10.2 Air quality at selected sites in Bulgaria (annual averages, micrograms/cubic metre)

City	Particulates		Sulfur dioxide		Lead	
	1990	1993	1990	1993	1990	1993
Assenovgrad	344	180	252	113	2.33	0.52
Bourgas	88	101	36	34	–	0.2
Dimitrovgrad	380	216	71	86	0.9	–
Galabovo	125	224	72	117	0.11	0.1
Kremitkovtsi	118	126	37	26	0.51	0.39
Kurdzhali	327	169	89	130	1.1	1.1
Pernik	179	387	122	44	0.48	–
Pirdop	453	208	330	206	–	–
Pleven	308	180	8	21	0.3	0.3
Plovdiv	485	360	35	88	1.5	0.7
Rousse	157	182	28	19	–	–
Sofia-Drouzhba	245	165	36	23	0.27	0.26
Svishtov	509	103	18	96	–	0.1
Varna	432	545	29	31	–	0.2
Veliko Turnovo	491	257	33	27	0.47	0.4
Zlatitsa	432	185	232	345	–	–

Source: Bulgarian environmental strategy study update, December 1994.

and the Maritsa thermal power plant which were among the first recipients of clean-up assistance from the European Bank for Reconstruction and Development.

The Bulgarian NGO sector in environmental clean-up

Having sketched in some necessary basic context, I now turn to consider the environmental NGO sector in Bulgaria. In this section I will evaluate Bulgaria's environmental NGOs in terms of their potential contribution to the development of civil society with reference to the four variables introduced in the opening section of the paper. I will talk first about the *increase in the number* of environmental NGOs, then about the *increase in the size and diversity* of the sector, then about the *legal and financial frameworks* within which they operate and finally about the *internal processes and decision making* that tend to typify these organizations.

The increase in the number of environmental NGOs

There may be currently as many as 206 environmental NGOs in Bulgaria according to recent censuses (Penchovska et al., 1993, 1997; Mindjov,

1996), ranging from small local level organizations concerned with very parochial issues, to larger groups operating at the national level and with more diverse sets of interests. This is a marked increase from the mere handful of groups that existed publicly at the beginning of 1989. An abridged version of Penchovska et al.'s (1997) authoritative listing is included as Appendix 10.1, from which the range and diversity of organizational forms and geographical locations are immediately apparent. However, mere recitations of censuses of NGOs may not be sufficiently enlightening. For example, environmental NGOs sometimes exist *on paper*, that is to say either legally through formal registration under the Family and Persons Act (discussed below) or through their inclusion in key listings or directories. In either case, they may well have become defunct. Working with Penchovska's 1993 and 1997 directories of Bulgarian environmental NGOs, I have found several examples of organizations that appear to have become defunct or whose basic data is inconsistently recorded. For example, it has proven difficult to independently verify the membership of the 'Green Patrols' Independent Society, as this organization's membership is split up into numerous regional groupings. Similarly, there are also examples of NGOs that have experienced a cyclical ebb and flow of activity; alternating between periods of dormancy and periods of intense activity. Usually these are related to the movement through the Bulgarian scene of key activists and organizers. In one example, the Foundation for Ecological Education and Training (FEOO) went through a period of dormancy caused by the departure of a key activist, but has since been reactivated after another individual stepped in to provide needed leadership.

Another difficulty occasioned by the fixation on the numbers of NGOs arises with the participation of prominent activists in *more than one* organization. In the Bourgas region, located along the Black Sea coast, there are several environmental NGOs which have been in operation since 1989–90, including SOS-Bourgas, Ecoglasnost-Bourgas, Bourgas Ecology Foundation, Bourgas Black Sea Club, and the Foundation Bourgas–Ecology–Man. These were originally formed and are now led by the same handful of individuals. At one point, for example, two individuals – a former mayor of the municipality (obshtina) and the director of the municipality's Environmental Directorate – between them had created or controlled virtually all the major environmental NGOs in the region. Though this intermixing of NGO memberships among activists is not of itself a problem – these local activists saw that their organizations are functionally differentiated – it does serve to give a false impression of the total volume of environmental activism, which after all is the object of the 'NGO-as-civil-society' argument.

Finally, fixation on the raw numbers of NGOs at a given point in time tend to hide the rather striking *geographies* of these organizations. As Appendix 10.1 clearly shows, the majority of registered environmental NGOs maintain official addresses in the five largest Bulgarian cities: Sofia, Plovdiv, Varna, Rousse and Bourgas; over 75 per cent of them in the capital city of Sofia alone. From a regional perspective, Bulgaria (along with Albania) has significantly fewer NGOs active in areas outside the capital than other countries in Central and Eastern Europe. Staddon (1996) has argued that the urban bias of these groups is important, as it tends to impose a particular perspective on environmental problems and on the relationships between policy, funding and action. There has been an observed tendency for this perspective to become hegemonic in direct relation of the increasing rootedness in the capital. Indeed, the abilities of these urban-based organizations to monopolize access to sources of foreign funding and aid further eclipses attempts at grass-roots organization in 'the provinces'. In this way the mediating role developed by many capital-based environmental NGOs becomes a hegemonizing one. This marked urbanization can have equally strong effects on the constitution of understanding of environmental issues, squeezing out peripheral positions. I will take up this important point in a later section of this chapter.

The number and diversity of participants in Bulgarian NGOs

The postulated strong link between growth of environmental NGOs and civil society in Bulgaria implies that there will be a concomitant increase in the number and diversity of individuals involved in these exercises in popular self-organization. After all, the civil society literature unfailingly assumes that access and actual participation in grass-roots organizations need to be completely open, at least in principle if not in practice. Consequently, with each passing year NGOs in Central Eastern Europe ought to be more numerous, have bigger and more diverse memberships and be involved in an ever-wider range of activities. One anticipates, for example, an increase in rural participation in environmental action; a factor whose relative absence has already been noted. But one would also anticipate increased participation by people of all ages and all ethnic, religious and cultural backgrounds. Precisely to the degree that such expressions of diversity can be empirically demonstrated, the neoliberal model of NGOs as civil society can be upheld.

To date however, most Bulgarian environmental organizations continue to have quite small membership bases. Data from recent surveys of environmental NGOs in Bulgaria suggest that fewer than 20 per cent of such organizations have more than 150 members with the median being closer to 25. REC's most recent survey suggests that 35 per cent have between 10 and 25 members, while the bulk of the rest have fewer than 100. A good many environmental organizations are even smaller still, constituted around the activities of perhaps half a dozen activists. Borrowed Nature Association is a good example, being a highly active and public group with less than 20 official members. While some NGO leaders identify this as a problem and have undertaken various efforts to attract new members, others are not concerned with the small size of their organization. In the former camp is the leader of *Ecoglasnost* and *Bourgas–Ecology–Man* in Bourgas, a city of 250 000 people located on the Black Sea coast. An activist of long standing, he has repeatedly identified low membership and indeed low public interest overall, as the most formidable challenge to the existence of his group. Combating what he believes to be a deleterious public apathy, this activist has instituted regular broadsheets describing environmental conditions in the Bourgas region (periodic updates of information originally contained in the 1991 *Black Book of Bourgas*), frequent columns on environment issues in local papers, numerous open meetings and school-based education programmes. Other groups are similarly concerned about lack of public engagement. During the summer of 1995, Green Balkans-Plovdiv travelled to several areas of Bulgaria with a photo exhibit detailing its activities and the condition of the Bulgarian environment in order to increase public awareness of these activities and generate additional support, particularly in regions of the country where it lacked membership.

Certainly a major factor affecting the size of environmental NGOs in Bulgaria is the severe economic crisis that has convulsed the country since 1989. Earlier in this chapter I touched upon the significant economic crisis currently being experienced in the country. Under these conditions, there is a potential for volunteerism to be hampered as people struggle for survival (Nikolov, 1992). Instead, it can reasonably be anticipated that people will choose to become involved only under specific and highly personal conditions: if they are in a financial or professional position that allows them to do so; if they are able to generate funding for projects in which they are interested by conducting the project under the auspices of an NGO; if their health or survival is affected by the issue being addressed by NGO activity (for example, in the author's detailed research on conflict over water resources, or the Rousse mothers joining together to protest against hazardous air quality); or if they are unemployed students or recent university graduates who can gain needed vocational experience and perhaps a bit of income by working with an NGO on an environmental project. I note here that this result mirrors results of studies of the NGO sector in other parts of the world (Wolch, 1989; Atampugre, 1997). Thus, the current economic situation affects both the size of the organizations and the kinds of people who are involved in them. And while this observation is not necessarily damaging to the neoliberal argument, it does unavoidably shift our attention away from the mere juridical possibility of membership towards the *material conditions* under which it is likely to come to fruition.

A clear conclusion emerging from this discussion is that there are significant problems with the postulated 'NGO-as-civil society' linkage. Indeed, I have already hinted that it may be misleading to pin the existence of something called civil society on the relative health of what are after all primarily *juridical subjects*, creatures of legal designation. Surely the potency of civil society must lie somewhere else, for example in the opening-up of a public space within which free associational activity is not just possible, but *practised*. This restatement of the problem points in three directions that will be explored in the following sections of the chapter. First, it is necessary to review the precise structure of law and state regulation to test the condition that NGO activities must be as free as possible from state intervention. Second, it is necessary to take a closer look at typical decision-making processes within environmental NGOs to ensure that they are internally democratic and open. Third, it is critical to observe whether environmental NGOs are popularly perceived as legitimate representatives of environmental interests.

The legal status of Bulgarian NGOs

The extent to which NGOs can be embedded in society and the way in which they carry out their activities is also affected by the institutional context in which they operate. This includes such things as whether they have the legal right to exist, what kinds of rights and responsibilities they have, their ability to intervene in the policy-making process, and to manage fund-raising activities. Bulgarian NGOs currently find themselves in a very complex and ambiguous relationship with national and international state structures. On the one hand, the antiquity of current laws and the real interpenetration of the state and non-state sectors in practice compromise the neoliberal ideal of the NGO as a juridically independent 'individual'. On the other, the current economic squeeze and the general lack of a culture of philanthropy in Bulgaria has meant that a great many NGOs are dependent upon foreign donors. Both situations work to bind NGOs ever closer to the needs of the state at national and international levels – a problem that is certainly among the greatest challenges to the NGO sector as a whole. In this subsection I briefly sketch out the legal and fiscal systems within which environmental and other NGOs operate.

The new Bulgarian constitution, adopted in 1991, guarantees citizens and groups some basic rights of public participation including the rights to free association, to a healthy environment, to information and to peaceful assembly. What this actually means in terms of the ability of NGOs to affect and participate in environmental decision making, particularly at the national level, is, however, less clear. There are no constitutional provisions or laws giving NGOs rights or guarantees to participation in legislative deliberation or the rule making of the parliament or government. The civil legal framework for Bulgarian NGOs is largely provided by LPF, the *Law on Persons and the Family (*1949*)*, which addresses such issues as the creation, registration, membership, governance, and dissolution of the organizations. This law specifies two kinds of organization, 'foundations' and 'associations', both of which have legal standing in the country following their required registration with the appropriate district court. The critical difference between them is whether or not they have members (associations do, foundations do not), although different provisions in the law apply depending upon how the organization is registered. For associations, in particular, the law includes regulations dealing with organizations' governance structure, requiring specifically a general assembly of members and a board of directors. That said, this law was originally adopted in 1949, has not seen significant revision in the post-communist period, and has been criticized as being ambiguous and outdated (Kyuthchukov, 1995).

Wholly missing from the LPF and the constitution are necessary clarifications about the rights of public access to information about and participation in the state policy-making process. The most complete public participation procedures are included in the environmental impact assessment provisions of the Environmental Protection Law, which grants rights to information, gives citizens a direct role in the environmental impact assessment process and to appealing administrative decisions. However, an April 1995 amendment to the law exempts from this requirement projects if 'the vital interest of the population is at stake'. There is also a general lack of precedent for the use of courts by NGOs to accomplish environmental objectives, and an incomplete set of legal tools for doing so. Given the limited opportunities for direct NGO participation in environmental decision making at the national level, it is perhaps not surprising that Bulgarian environmental NGOs often resort to various non-formal techniques such as organized street protests, contacting the media, petition campaigns, and efforts to influence the political process by issuing communiqués to the international community.

Internal decision making: the democratic deficit

The analysis so far suggests that there may be something of a 'representative' or 'democratic' deficit in the environmental NGO sector. With few exceptions, memberships are quite low, agendas are set by small expert core groups, and Western funding bodies tend to insist on a tightly delimited project oriented mode of organization (Sampson, 1996). This problem of a democratic deficit is quite important from the point of view of Western theorizations of the links between the growth and proliferation of NGOs and civil society. After all, if it can be shown that even a sizeable minority of these organizations are in fact nothing more than 'little empires', then their status as expressions of a re-emergent civil society must surely be rendered highly questionable. Investigation of these issues requires more than relatively simple enumeration and macro-economic analysis. In this section, I provide some results from ethnographic study of Bulgarian environmental NGOs with particular attention paid to internal decision-making processes and to the strategies employed for the constitution and representation of environmental problems.

All of the above observations about the character and activities of Bulgarian environmental organizations carry with them the implication that leadership structures are oriented towards the top-down, expert-led model, rather than the bottom-up, consensus-driven model. Certainly this is true for many of the Bulgarian environmental NGOs examined by the authors. Undoubtedly this is at least partly a consequence of the small membership sizes, though it is also often the product of an *expert-led and somewhat autarkic approach* to organization. This is reflected in the surprisingly low public profile of even the largest and 'objectively' most important of these groups. Contrary to expectations generated by the neoliberal model, environmental NGOs in Bulgaria continue to suffer from a relative lack of public visibility. Social surveys by Pickles and Staddon (1994a,b) reveal that individual knowledge of environmental organizations is very slight, as is the estimation of their potential efficacy in addressing environmental problems. As in Western countries, there is a strong tendency within the NGO sector for activities to be directed by and identified with a small number of highly active figures. Thus Ecoglasnost-National Movement is overwhelmingly identified with Edvin Sugarev, Green Patrols is identified with Amadeus Krastev and the Green Party with Petr Slabakov. Interestingly, public opinion polls have repeatedly shown that the organizations themselves have a much reduced public profile compared with their well-known leaders, and can sink into obscurity should these key figures move on. The cyclical ebb and flow of NGO fortunes that this can create has already been noted, as has the generally low level of public recognition of the groups themselves, apart from their key figures.

Another problem, when focusing on presumably the non-governmental character of civil society, is that it is not uncommon for these organizations, particularly the Sofia-based ones, to have close ties to government ministries, universities and research institutes by virtue of the fact that the NGO leaders were or still are employed by these organizations. The key reference groups for organizations such as Ecoglasnost-Movement and Borrowed Nature are not Bulgarian communities or even the general public, but rather the environmental ministries of the central government and of international groups. These urban-based professionals may be unfamiliar with the rural social and economic contexts in which conservation or natural resource management activities take place, thus limiting their ability to interact successfully with local communities. Some observers have sought to justify this core-oriented technocratic focus in terms of the claim that 'Bulgarians are not by nature co-operative and consultative ... a lot of learning is needed' (Staddon, 1996). Whatever the truth of such assertions, for practical purposes it is the contacts with government agencies and foreign NGOs that lend practical support and formal political legitimacy to these Bulgarian environmental groups, rather than any putative connections with local communities. In other words, the successful achievement of environmental projects by Bulgarian NGOs often seems to depend more on 'horizontal' connections with national and international state apparatus, than it does on 'vertical' connections with communities and local publics.

Towards an environmental research agenda

Of course, there is a great more that I could say about Bulgarian environmental NGOs; interested readers are directed to the 'further readings' list at the end of this chapter. I hope, however, that I have established sufficiently that the environmental NGO sector in Central-Eastern Europe is far too complex to neatly fit into the 'NGO as civil society' model. Indeed the NGO sector, troubled as it is by problems of legal ambiguity, focus, co-operation and funding, is far richer and more varied than much of the scholarly literature has previously allowed. So too, the idea of 'civil society', as an essentially contested concept, is too complex and even ambiguous to admit of any easy operationalizations. Consequently the simplistic 'NGOs as civil society' model cannot be accepted, and we must seek to construct more sophisticated interpretations of NGO sector development during post-communist transition. This is an as yet largely unaddressed research frontier (but see Creed and Wedel, 1997; Cellarius, 1998).

In these concluding comments I would like to reflect upon how these lines of analysis and research can contribute to an emerging environmental research agenda that is interdisciplinary in nature, theoretically informed and empirically grounded. One of the things that I am trying to do as part of a new research programme is search for useful comparisons with other transition countries. Current planned research involves examining local environmental management initiatives in Poland and perhaps also the former Soviet Republic of Georgia. The focus here is not on environmental NGOs *per se*, as I am more interested in the optic of civil society as a way of possibly looking at the ways in which localities, especially peripheral ones, are reconceiving of their local environments as resources for community development. Of course this may involve some work with local NGOs, as recognized in law, but I think that it is important not to prejudice the enterprise by preordaining the sorts of organizational form sought – this sort of tautology has occasionally surfaced in the NGO literature already. Research during the summer of 1997 in rural Bulgarian localities revealed that there are numerous social (extended family land tenure patterns), cultural (reconceptions of the relations between in-group identity and space), economic (communal use of local lands) and other socio-cultural forms (re)asserting themselves that may have important implications for environmental management (cf. Staddon, 1998). For this reason, my new research programme focuses on 'local environmental initiatives', rather than NGOs as such.

Using anthropological perspectives, my colleague Barbara Cellarius has developed quite detailed ethnographic knowledge about the complex ways in which local knowledges are (re)asserting themselves. She has also demonstrated quite clearly that relations between environmental NGOs and local communities are not always smooth even when both are ostensibly committed to conservation (Cellarius, 1998). There is certainly scope for the elaboration of such perspectives in other transition localities, a judgement fully recognized by Creed and Wedel (1997). In a related vein, the *sociology* of environmental NGOs is worthy of more attention. This chapter has presented some sociology material, particularly in the sections on the internal characteristics of NGOs and their decision-making processes, but more needs to be done. Related work in the sociology of NGO participation could provide useful insights about how and why people work within NGOs. This chapter has suggested some of the more instrumental reasons for NGO participation, but surely these are not the only motivating factors.

On the more technical side, there is a considerable gap in the research on the contributions that environmental NGOs have made with respect to specific technical or managerial issues such as choice of water filtration technologies, database management protocols, and so on. Engineers of all stripes have a considerable amount of practical experience working with and through NGOs, and it is to be hoped that their experiences can find their way into the scholarly debate about the role(s) of environmental NGOs in post-communist transition. Among the few such interventions are the volumes of Goldsmith and Hildyard (1984) on large dam developments, which at points are quite critical of international and local environmental NGOs in developing world contexts.

Acknowledgements

I would like to thank Barbara Cellarius of the Department of Anthropology at the University of Kentucky for permitting me to adapt material from a forthcoming joint publication for this talk. I am thankful also for encouragement and assistance I have received from Prof. John Pickles also of the University of Kentucky, Gavin Bridge of the University of Oklahoma and Richard Spalding of the University of the West of England.

The full paper, 'Non-governmental Organisations, the Environment and Civil Society in Bulgaria', jointly authored with Barbara Cellarius of the Department of Anthropology of the University of Kentucky, will appear in *East European Politics and Societies*, 16(1). I am grateful to Barbara for permission to use some of the material from that paper here.

Appendix 1: Selected environmental NGOs in Bulgaria, 1998

Name	No. of members	Year founded	City/town based in
Academic Youth Environmental Club 'Amek'	58	1990	Sofia
Association Green Balkans – Plovdiv	190	1990	Plovdiv
Birds of Prey Protection Society	300	1990	Sofia
Black Sea Club – Bourgas	–	1993	Bourgas
Borrowed Nature	20	1992	Sofia
Bourgas–Ecology–Man Foundation	–	1992	Bourgas
Bulgarian Association of Experts in Park and Landscape Architecture	–	1992	Sofia
Bulgarian Botanical Society	200	1823	Sofia

Name	No. of members	Year founded	City/town based in
Bulgarian Ecologists Association (Abekol)	80	1991	Sofia
Bulgarian Environmental Culture Club	180	1991	Sofia
Bulgarian Geographical Society	1000	1923	Sofia
Bulgarian National Association on Water Quality	–	1994	Sofia
Bulgarian Nature Scientists Society	–	1896	Sofia
Bulgarian Society for Animal Protection	200	1989	Sofia
Bulgarian Society for Conservation of the Rhodope Mountains	4600	1990	Chepelare
Bulgarian Society for the Protection of Birds	594	1988	Sofia
Child and Nature Foundation	–	1991	Sofia
Earth and Man Foundation	50	1993	Sofia
Earth and Resources Foundation	–	1997	Sadovo (Plovdiv)
Ecoimmunoprophylaxis Center, Ecoimmunodeficiency Control Foundation	–	1991	Sofia
Ecological Society 'Ekoexpert'	–	1996	Plovdiv
Ecomission Club	–	1997	Varna
Ecomonitoring Club	10	1992	Sofia
Eco-Sphere Foundation	–	1992	Vidin
Environmental Protection Club for University Students – Higher Forestry Institute	50	1977	Sofia
Environmental Training and Education Foundation	–	1991	Sofia
Green Patrols	4500	1991	Sofia
Independent Association for the Ecological Protection of the River Vit	300	1991	Pleven
Independent Club of Experts 'Green Future'	50	1992	Sofia
National Children's Ecological Movement 'Don't Pollute'	500	1994	Sofia
National Movement of Ecoglasnost	3000	1989	Sofia
National Partnership Forum for Sustainable Development	–	1997	Sofia
Nature Protection Society	100	1987	Razgrad
Public Environmental Centre for Sustainable Development	–	1996	Varna
Scouts Tourist Club – Gabrovo	30	1996	Gabrovo
United Front for the Ecological Salvation of Kurdzhali	2700	1990	Kurdzhali
Wilderness Fund	80	1990	Sofia

Sources: Penchovska, Ivanova, Kobakova (1993); Penchovska, Petrov, Kobakova (1997); Mindjov (1996); Shoumkova et al. (1998).

References

Atampugre, N. (1997) 'Aid, NGOs and grassroots development'. *Review of African Political Economy*, 71, 57–73.

Bernhard, M. (1993) 'Civil society and the democratic yransition in East Central Europe'. *Political Science Quarterly*, 108(2), 307–26.

Cellarius, B.A. (1998) 'Linking global priorities and local realities: nongovernmental organizations and the conservation of nature in Bulgaria'. *Bulgaria in Transition: Environmental Consequences of Political and Economic Transformation*, pp. 57–82. Ashgate Press.

Cellarius, B. & Staddon, C. (2002) 'Environmental non-governmental organisations, civil society and democratisation in Bulgaria'. *Eastern European Politics and Societies*, 16(1).

Cohen, J. & Arato, A. (1990) *Civil Society and Political Theory*. MIT Press.

Creed, G.W. & Wedel, J.R. (1997) 'Second thoughts from the second world: interpreting aid in post-communist Eastern Europe'. *Human Organisation*, 56(3), 253–64.

Diamond, M. & Plattner, S. (1992) *The Global Resurgence of Democracy*. The Johns Hopkins University Press.

Goldsmith, E. & Hildyard, N. (1984) *The Social and Environmental Impacts of Large Dams. 2 volumes*. San Francisco: Sierra Club.

Jancar-Webster, B. (1993) 'The East European environmental movement and the transformation of East European society'. In B. Jancar-Webster (ed.), *Environmental Action in Eastern Europe: Responses to Crisis*, pp. 193–219. ME Sharpe Press.

Keane, J. (1988) *Democracy and Civil Society*. Verso.

Kligman, G. (1990) 'Reclaiming the public: a reflection on creating civil society in Romania'. *East European Politics and Societies*, 4(3), 393–427.

Kyuthchukov, S. (1995) 'Bulgaria: country report'. In D.B. Rutzen (ed.), *Selected Legislative Texts and Commentaries on Central and East European Non-for-Profit Law*. Sofia, Bulgaria: International Centre for Non-for-Profit Law, European foundation Centre, and Union of Bulgarian Foundations.

Mindjov, K. (1996) *Bulgarian NGOs: Towards the NGO Parallel Conference 'Environmental for Europe'*. Sofia, Bulgaria: Borrowed Nature Association.

Nikolov, S. (1992) 'The emerging nonprofit sector in Bulgaria: its historical dimensions'. In K. McCarthy et al., *The Nonprofit Sector in the Global Cummunity: Voices from Many Nations*, pp. 333–48. San Francisco, CA: Jossey Bass Publishers.

Paskaleva, K., Shapira, P., Pickles, J. & Koulov, B. (1998) *Bulgaria in Transition: Environmental Consequences of Political and Economic Transformation*. Ashgate Press.

Penchovska, J., Ivanova, M. & Kobakova, A. (1993) *Catalogue of Environmental Non-governmental Organisations in Bulgaria*. Sofia: Regional Environmental Centre.

Penchovska, J., Petrov, D. & Kobakova, A. (1997) *Spravochnik na nepravitelstvenite organizatsii za opazvane na okolnata sreda (Catalogue of environmental nongovernmental organizations in Bulgaria)*. Sofia, Bulgaria.

Pickles, J. & Staddon, C. (1994a) *Perceptions of Environment, Well-Being, and Governance in Bourgas and Kameno Obshtini in June 1992*. Research Paper #9417. West Virginia University: Regional Research Institute.

Pickles, J. & Staddon, C. (1994b) *Attitudes About Governance and Environmental Issues in Bourgas and Kameno Obshtini*. Research Paper #9413. West Virginia University: Regional Research Institute.

Sampson, S. (1996) 'The social life of projects: importing civil society to Albania'. In C.M. Hann & E. Dunn (eds), *Civil Society: Challenging Western Models*, pp. 121–42. London: Routledge.

Seligman, A. (1992) *The Idea of Civil Society*. Princeton University Press.

Shoumkova, T. et al. (1998) *Who is Doing What for the Environment in Bulgaria*. Sofia, Bulgaria: Borrowed Nature Association.

Snavely, K. & Desai, U. (1995) 'Bulgaria's non-profit: the search for form, purpose, legitimacy'. *Voluntas*, 6(1), 23–38.

Staddon, C. (1996) *Democratisation, environmental management and the production of the new political geographies in Bulgaria: A Case Study of the 1994–95 Sofia Water Crisis*. Unpublished Ph.D. dissertation. University of Kentucky: Department of Geography.

Staddon, C. (1998) '*Localities, Natural Resources and Transition in Eastern Europe*'. Presented at the Annual Meetings of the Institute for British Geographers, Kingston University, 5–8 January 1998.

Walzer, M. (1992) 'The civil society argument'. In C. Mouffe (ed.), *Dimensions of Radical Democracy*, pp. 89–107. Verso Press.

Wolch, J. (1989) 'The shadow sector: how territory shapes social life'. In M. Dear & J. Wolch (eds), *The Power of Geography*, pp. 197–221. Unwin Hyman.

World Bank (1994) *Bulgarian Environmental Strategy Study: Update and Follow-up*. World Bank.

Wyzan, M. (1996) 'Stabilisation and anti-inflationary policy'. In I. Zloch-Christy (ed.), *Bulgaria in a Time of Change*, pp. 77–106. Avebury Press.

Yancey, N. & Siegel, D. (1994) 'The nonprofit sector in East Central Europe'. *Transnational Associations*, 1, 23–24.

11
Environment and Collective Action Under Threat: Tales from the Hindukush

Geof Wood

Introduction

Neoliberal globalization has over-privileged markets and individualism as the way in which the interaction between livelihoods and natural resources are managed. Sustainability of the environment and development is to be achieved through rationing via price, the presumption of a calculus in which as scarcity becomes more evident the economic pressure for conservation increases as the costs of depletion are internalized. However, the magic of the market to deliver sustainable outcomes (the invisible performance of rationality), relies upon assumptions of immobility with respect to space and resource-use dependency, neither of which reflects the reality of mobility and resource hopping in which capital resembles the locust rampaging through the grain fields. Global markets allow for resource hopping, relying upon continuous technological innovation to overcome the constraints of previously depleted technologies and associated resources. It is thus hard to argue that global markets are compatible with sustainability. But what happens where global markets are weaker, where the condition of immobility is much stronger but nevertheless threatened by the carrying capacity of the local environment interacting with population growth and the presence of exit options? What happens when a local population experiences a paradigm shift of principles and values by which the interaction between livelihoods and key local resources is organized?

This is a general question which can still be applied to many parts of the world, especially the poorer, remoter parts. Some might argue that this is now merely a 'transitional' question of residual historical interest as global integration continues apace, swamping and homogenizing local resource management systems. Modernization and hegemony of markets will replace local, 'indigenous', pre-capitalist, community based systems of management, albeit with some aspects of national and international regulation, as for example in Kyoto. Others would dispute this account of history and future institutional trends, not as a matter of prescription but of fact. Long has written of globalization and localization (1994), and also of re-localization. It is the plea to understand not just diversity but the logic of its intensification within a refined comprehension of what globalization really means: socio-economic post-modernism, in other words.

To illustrate this logic of diversity, this paper examines some behaviour in one of the remotest corners of the world, the high mountain areas of Hindukush and the Karakoram in Northern Pakistan, with a population of approximately 1.3 million. An area reluctantly and ironically in the news as I write, and perhaps when this is published too. Although remote in the sense of being precariously linked to contemporary global markets, these areas have always been 'connected' via their location on key trade routes. My own exposure to these areas has been extensive from October 1993: through a short stay in Booni village, upper Chitral, North West Frontier Province (NWFP; or Badakshan, if considering older, pre-Raj identities which mildly persist) supervising a research student; through fieldwork in 1996, while reviewing the work of the Aga Khan Rural Support Programme (AKRSP) (Wood, 1996), across the Hindukush of Chitral and the Northern (Karakoram) Areas of Gilgit, Hunza and Baltistan; and through further continuous pieces of fieldwork from 1997 to date in the same areas, working alongside staff in the AKRSP in household, village and thematic studies.

Location and characteristics

The area has to be understood as a part of the mountain desert systems of Central Asia, with: extreme variations in summer and winter temperatures; significant light and shade variation, affecting photo-periodic cropping in terms of heat, length of day and length of season; very low levels of direct precipitation onto the alluvial fans of the valley bottoms; regular winter snowfall at higher altitudes; a large concentration of the world's glaciers; mixed agriculture, horticulture and livestock farming systems, concentrated into the alluvial fans; net deforestation, after afforestation is accounted for; fragile high summer pastures and constrained lower altitude grazing; and essential irrigation provided by surface water channels, sourced from glacial melt. It is an area of gentle rather than steep terracing, since the steep hillsides are quickly barren.

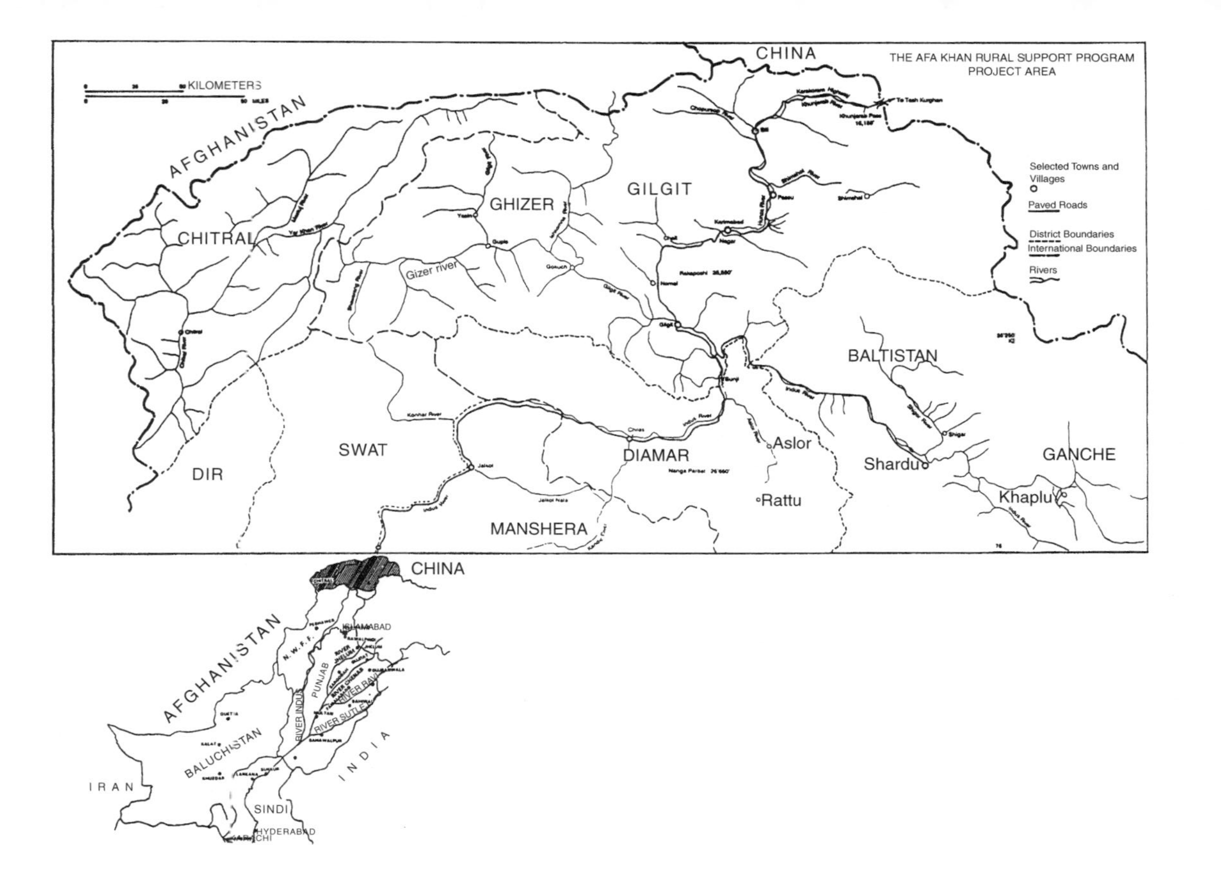
THE AFA KHAN RURAL SUPPORT PROGRAM
PROJECT AREA
KILOMETERS
Selected Towns and Villages
Paved Roads
District Boundaries
International Boundaries
Rivers
CHINA
AFGHANISTAN
CHITRAL
GHIZER
GILGIT
Gizer river
BALTISTAN
DIR
SWAT
DIAMAR
Aslor
Shardu
GANCHE
Rattu
Khaplu
MANSHERA
CHINA
AFGHANISTAN
PUNJAB
RIVER INDUS
RIVER SUTLEJ
BALUCHISTAN
INDIA
IRAN
SINDI
HYDERABAD

Dwellings tend to be clustered for mutual protection, with some contemporary evidence of residential overspill onto erstwhile cropping land. It is therefore semi-pastoral, with a high, though declining, dependency upon natural resource management, especially water.

Until two decades ago, these communities were almost entirely dependent upon local natural resources, with some element of trading in order to acquire non-local goods.

History and settlement

The three main areas in this Hindukush–Karakoram (HK) region (Chitral district of NWFP, Ghizer–Gilgit–Hunza, and Baltistan – usually referred to as the Northern Areas and Chitral [NAC]) – were princely states surviving beyond the partition of British India into the mid-1970s. This is the arena of the Great Game, in which the British deliberately created a mountainous bulwark to defend its Indian empire against Russian incursion. Political Agents were posted from the late-nineteenth century offering a combination of stick and carrot to local rulers to ensure adequate loyalty to turn them away from Russian and Chinese suzerainty. The Wakki clan territory was allocated by the British and Russians to Afghanistan as the Wakkhan Corridor (running west–east to the source of the Oxus across the 'top' of the Chitral and Ghizer valleys) to act as a buffer between the two nineteenth-century superpowers. This now represents 'problematic' territory between Pakistan and Tajikistan, held at the time of writing by the non-Pashtun 'Northern Alliance' clans in contrast to the Pashtun tribes, south of the Panjshir valley (that is, the Taliban during 1996–2001). The erstwhile ruling families in the different valleys retain some influence in their respective areas, though perhaps more so among the Mehtars of Chitral than elsewhere. But even here in the Mastuj sub-division of Chitral, recent Union elections did not return the son of the prominent noble. Since the areas are almost entirely populated by Muslims from different sects, they opted for Pakistan at the time of Partition in 1947, with the eastern and southern boundaries of Baltistan representing the UN line of control with Indian occupied Kashmir. Many of the trade and cultural routes though to Ladakh and Srinagar were thus severed, as the area had to reorient its communications through the Indus valley to the downcountry Pakistani cities of Islamabad (the new capital created from 1967), its neighbour Rawalpindi, Lahore, and, for Chitral, Peshawar. Connections from all parts of the HK region to the commercial capital of Karachi on the southern coast are significant for education and employment.

Since the Baltistan and Ghizer–Gilgit–Hunza areas are considered to be part of disputed Kashmir, successive Pakistan governments have been reluctant to absorb these areas into mainstream national politics through fear of weakening, *de facto*, their demand for a plebiscite for the whole of Kashmir with the three options of independence, opting for India or opting for Pakistan on the agenda. As a result, these areas, collectively known as the Northern Areas, have special governance arrangements outside the federal system (that is, local councils with limited powers, an appointed Minister rather than democratic representation at national level, local rule by civil officials) and a high army presence operating continuous, quasi martial law. Although the population of the Northern Areas is entirely Muslim, three main sects are represented plus smaller ones. Thus Baltistan is mainly populated by Shia, with some Noorbakshi, and since the Iranian revolution of 1979, has even been nicknamed as 'little Iran'. There are also significant connections to the Shia of Kuwait. The 100 mile long Hunza valley is mainly populated by Ismaili, who have brethren scattered in Gilgit, Ghizer and upper Chitral, as well as in the central Asian, 'Badakshani', states to the north. But there are also Sunni in these areas too, with some Shia in lower Gilgit.

Chitral to the west, by contrast, is formally integrated into the federal system as a district of the NWFP, and returns members to the Provincial Assembly and the National Assembly (when they have not been suspended by martial law, as at the time of writing). However, the area also faces uncertainty and insecurity, derived partly from the instability of its neighbour Afghanistan and the consequent influx of refugees; partly as a cultural, ethnic and religious frontier between Pashtun (or Pathan) and Chitrali, and between Sunni and Ismaili sects, with the Sunni concentrated in the lower half of the district and the Ismaili in the upper half (though with some Sunni dominated valleys like Turkho); and partly because it is separated from its nearest southern neighbouring district, Dir (strongly, even fundamentally, Sunni), and through it access to down-country and Peshawar by the 10000 feet Lowari Pass, which is impassable during the six or seven months of winter. Indeed, during these months, the only road access is through the Kunar valley which runs into Afghanistan before reconnecting to Pakistan via the Kyber Pass.

Overlaid across these broad ethnic and sectarian differences in the NAC region is a complex kinship system in which families are essentially grouped into identifiable clans. While they are broadly coterminous with sectarian divisions, there are instances in which clans might straddle sects due to sporadic conversions and inter-sect marriage. Again these clans are mainly endogamous, but not exclusively so. Generally within Muslim society, marriage within small kinship units is quite common including cross-cousin marriage, thus reinforcing a general practice of clan endogamy. A large village will typically be divided into mohallahs (each mohallah with its own religious centre) which are most probably of a single sect (where the village is a mixed sect). In single sect villages, these mohallah may themselves be coterminous with a clan but even this is difficult to generalize with different clans from the same sect living closely together. A village may typically comprise 5–10 distinct clans. Historically, these clans may have been associated with particular occupations, thus resembling a caste/jati structure though with a much reduced sense of pollution: ritual or real. Currently these occupational distinctions have receded, although in some areas some clans retain a higher political and social status than others, derived either from being part of the ruling clan (for example, Kators in Chitral as the clan name of the ruling Mehtar family) or being favoured with administrative office by the ruling families. Such clans retain a certain dominance in village affairs, partly based on lingering respect, partly based on superior assets arising from this ruler association, and partly due to inter-linked superior education with some clan members holding secure, professional jobs.

There is a further linguistic classification to be thrown over the NAC region. The primary language for most inhabitants will be quite localized. Most people in Chitral speak Khowar, sometimes referred to as Chitrali, regardless of sect, clan and location. Thus despite the sharp Sunni/Ismaili contrast in settlement between lower and upper Chitral respectively, the virtual common language of Khowar has been cited as an important ingredient of cohesion and sense of Chitrali identity – even to the point of nationhood, though that is probably far-fetched. In a sense, language and sect compete for identity, with a fear among the fervent Chitrali that sectarian division will win with the Chitrali Sunni increasingly identifying with their sectarian brethren in the fundamentalist Dir District to the south over the Lowari Pass. Elsewhere in Ghizer–Gilgit–Hunza there is Khowar (in Ghizer) shading into Shina as we move east to Gilgit, Shina in lower Hunza–Nagar shading into Burushuski as we move north up through the valley, itself shading into Wakki as one reaches the Gojal area of upper Hunza. This Wakki speaking area thus connects to the extreme upper part of Chitral as one tracks back west from upper Hunza inside the Pakistan side of the Wakkan corridor. In Baltistan, there is again the virtual unifying language of Balti spoken, coterminous with the Shia sect to which more than 90 per cent of the population belong. These distinct linguistic regions have to be understood as primarily geographical in explanation rather than reflecting social or religious categories. For example, depending precisely where you live in Ghizer, regardless of being Sunni or Ismaili in this mixed sect area, you will either have Khowar or Shina as your first language. Likewise, Shia Nagar and Ismaili Hunza will both speak Shina at the lower, southern end of the valley and Burushuski further to the north. For all of these groups, the more mobile family members due to status, education and other necessary migration will have some knowledge of neighbouring languages. And if their movement is further afield, then many will share Urdu as a second language, and for the top elite even English. Inherently there is a strongly gendered pattern to familiarity with other than a primary language, as well as reading and writing.

Before the reader becomes over-frustrated with this, still superficial, ethnographical tour of the region, there are some preliminary conclusions to be drawn which impact upon the main argument below. The 'frontier' characteristics of the NAC region have constituted a history of dynamic settlement reflecting: invasion (the genetic outcomes of the presence of Alexander the Great in the region can still be seen); trade movement; local migration due to squabbles or pressure on fragile resources in nearby areas; and induced migration by order of ruling families to settle new alluvial fans and thereby bring them under control for revenue purposes as well as extending political territory in competition with others. In most villages of my fieldwork, informants will invariably trace their ancestry to somewhere else within the region. Combine this history of settlement with the topography, in which valleys and sub-regions can be relatively inaccessible from each other, and the socio-cultural heterogeneity, and diversity can be appreciated. With a slow but steady increase in communication via road links, some of this diversity might reduce (to be discussed below). However, with this settlement legacy, we encounter an area characterized by cross-cutting ties and fission–fusion as the basis of order and disorder, in which people simultaneously inhabitant parallel, contingent communities, variously active or passive according to circumstance, within a framework of feudal authority which has set limits to local conflict as the basis for competition between larger, aggregate socio-political units. Thus we have a social basis for collective action which is partially induced by feudal authority, and partially underpinned by the principle of contingent community (see Streefland et al., 1995 for further elaboration of some of the above description, though most of what is presented here is based on my own primary fieldwork).

Feudalism and its demise

The feudal system of the princely states persisted after Partition up to the early 1970s and its formal abolition under the People's Party government of Zulfikar Ali Bhutto. Of course remnants of status, property and respect remain, with some members of erstwhile ruling families moving into politics alongside business activity and the holding of other offices. Before examining the significance of abolition, two essential characteristics of the feudal system are highly relevant to the analysis which follows. *First,* the term 'feudal' is properly used from its European context of control over persons rather than control over land (the term has been frequently misused in other South Asian contexts to describe the latter). Control over persons reflects earlier conditions under which labour was the scarcest factor of production, even in these mountain areas where cultivable land was only a minor fraction of the whole. In any epoch, the dominant mode of production will always be centred around control and management of the scarcest means of production. The political subjects of ruling families would have their movements curtailed, with some valley 'frontier' passes even policed to prevent escape (the origin anywhere of passports and exit visas) or to charge taxes (significantly in kind – that is, substituting the labour of other family members) for the privilege (that is, for trade, education or other purposes). Political subjects might also be posted to settle other villages (as noted above). *Second,* such control over persons was significant for the creation of public goods under low technology, low efficiency conditions: creation and maintenance of irrigation channels; road and paths; bridges; terracing of alluvial fans; and of course construction of forts and religious buildings. Given the high dependency of all upon irrigation under mountain desert conditions, the feudal mobilization of labour for this key productive good has been described as 'hydraulic society'. The term has broader application in Central, South, East and South-East Asia where both canal and tank irrigation has required similar labour mobilization beyond immediate social space. In other words, where hydraulic space transcends social space, so principles of wider authority have to transcend local collective action based upon a sense of local community, in order to overcome the fission between such communities. Such systems have also been referred to as 'oriental despotism' or the 'Asiatic mode of production'. The main challenge to the inhabitants of these societies was how to maintain and create expanded public goods, especially the productive ones and ones coming under threat such as forests and pastures, while these feudal institutions were being abolished throughout the nation-state of Pakistan, including the non-federal Northern Areas. However, the timing of this challenge has coincided with other changes which have brought the market in various forms into the reckoning as a set of social institutions to compete with non-market principles of collective action.

New communications: the Karakoram Highway

The livelihood conditions in the Northern Areas especially have been significantly transformed by the completion of the Karakoram Highway (KKH) in 1978. By opening up an all weather route from Islamabad to Kashgar in the Xinjiang Province of Western China, the Gilgit and Hunza sub-regions have been directly linked to each other as well as downcountry, with Ghizer to the west and Baltistan to the east indirectly linked by local roads (not yet completely metalled and continually threatened by landslides and heavy snowfall at the higher altitudes). While it would be normally oversimplistic to attribute widespread socio-economic change to a single road, the KKH is a special case of extending communication options beyond the narrow elite using planes and second homes downcountry for lengthy seasonal migration. Travel was exceptionally arduous before the KKH. The effects of the KKH will therefore feature widely across the analysis which follows. For Chitral, there has been no equivalent event, despite much speculation about all weather routes and/or tunneling through the Lowari pass. However, summer/autumn communication with downcountry areas has been long established, so that wider options have been available, though only seasonally.

Other recent changes

With this coincidence of improved communications and the abolition of the feudal political structures from the 1970s, we are thus able to review significant change in livelihoods and institutions over a two-decade period. At the beginning of the twenty-first century, characterized as it is by forces of globalization, this represents a unique opportunity to test a thesis about the power of markets to integrate local institutions with distinctive non-market values and social practices. The HK region has a rapidly growing population (growth rates in excess of 3 per cent are reported (Streefland et al., 1995) and successive five yearly World Bank evaluations of AKRSP). Higher standards of living are expected from a natural resource base which is therefore declining in productive value per household. There is an intensification of land use, with a corresponding higher dependence upon irrigation and more vulnerability to erosion. There is reduced village grazing land as a consequence of intensifying agriculture with implications for pasture management. There is a reduction in the proportion of farm-sourced income, compensated by the expanded opportunities for diversifying sources of income. However this is also leading to increased socio-economic diversification and social differentiation within local communities. These conditions then lead to the central question: to what extent is the increasing pressure on the local natural resource base, threatening environmental sustainability, coinciding with increasing inequality, individualization and a consequent decline in a local capacity for collective action to manage such environmental scarcity? How far are we able to conclude that the circumstances of continued reliance upon the local natural resource base, particularly reinforced by national level political instability and economic recession, restricts the opportunities for resource hopping, confines livelihoods solutions to local space, and thus inhibits the progress of markets to define the principles of resource allocation while ensuring that any costs of depletion will be borne locally? If so, then the route to natural resource sustainability will be via a retention of forms of collective action rather than market internalization. However, a further complication is the need to disaggregate this kind of conclusion by both location and type of resource.

Let us therefore examine the contextual conditions, summarized above, in more detail. The area appears to be experiencing the classic demographic bulge with declining mortality rates (including infant mortality) not yet impacting upon fertility rates. Other more context specific factors are also at work: strong cultures of patriarchy reducing female autonomy to initiate contraception combined with a purdah complex (Sharma, 1980) reducing mobility to access clinics, which can anyway be topographically remote. With population growth as a poverty indicator, it is no surprise to find larger family size among poorer families, which are more dependent on local farming, having fewer non-farm options. The AKRSP Farm Household Income Survey (1999) reported farm income as a percentage of household income as 54 per cent for the top income quintile compared to 82 per cent for the lowest income quintile. This translates into a well-observed fragmentation of holdings through multiple inheritance and a consequent intensification of land use where possible (that is, in the lower altitude, double cropped zones).

Where such options do not exist in single cropped zones, more families are compelled to seek off-farm income. The resulting outmigration, predominantly male, can be classified in various ways: localized or downcountry; daily, seasonal or longer term; unskilled, low productivity, low return employment contrasted to its educated opposite; for immediate income solutions or for investment in the longer term via the education of sons, and daughters for some sects, localities and classes. Local outmigration, especially from single cropped, high altitude, more remote locations, has partially stimulated the rise of growth poles, which have been further aided by government investment in decentralization with relocation of officials and the accompanying rise of construction and other services. This is leading, over time, to some spatial restructuring of the population as younger adults leave elderly parents behind in the early phases of such migration. The male dominated outmigration in search of off-farm opportunities is also having the effect of 'promoting' women into farm decision-making and natural resource management. This trend is particularly significant for community level collective management given the cultural constraints to women representing their family interest in public, male dominated forums.

In some localities, ecologically appropriate to new cropping options, the need to increase land productivity has inaugurated a shift from subsistence to the commercial farming of potatoes. At the higher altitudes, for example at the northern, upper (Gojal) end of the Hunza valley, the production is for the higher valued seed potato, with the colder temperatures reducing the probability of virus infection. At lower altitudes, the production is for table potato (that is, direct consumption). The entire NAC area is considered by downcountry entrepreneurs to have an ecological comparative advantage in potato production due to altitude and (for potato) virgin soils. However it is only areas well connected to the KKH which can easily contemplate shifting to such production due to transportation access. Potato production has several implications. So far it is an individual rather than collective strategy, leaving individual farmers isolated in fixing prices for both produce and inputs. The necessity to interact with strangers to the family, culturally requires males rather than females to manage the crop. The early financial attractions of the crop tempt farmers into overcommitting their land (spatially and intensively), thus entailing early experience of declining yield/fertilizer coefficients.

In some other, selected, locations, there has been a steady though not yet dramatic increase in tourism. Clearly a range of local as well as distant, national and global conditions affect the strength of this tourism through determining the extent of security in the region. For example, the 'Clinton' cruise missiles into Afghanistan after the attack on the US embassies in Kenya and Tanzania in 1998; the killing of a Sunni mullah in Chitral in August 1999, provoking sectarian tension between Ismaili and Sunni communities; the Kargil war between Pakistan and India in June–July 1999; the testing of nuclear weapons in India, followed by Pakistan; the aftermath of 11 September 2001 involving both heightened tensions in Pakistan as well as increasing tensions with India over Kashmir and the prospect of renewed hostilities; and so on. However in certain pockets in Hunza and Skardu-Baltistan, the scale of tourism has been sufficient to have an impact on various aspects of environmental consumption – especially energy, transport, water and wildlife. The possibilities exist for an expansion of tourism in Chitral as well, if security conditions improve. The central problem with tourism in a region like this is the extent to which it offers the prospect of private gain at public cost, and the extent to which this private gain is substantially non-local (transport, hotels and services) while the public cost is born locally. Of course there is some local employment generation for individuals (hotels, guides, porters and local crafts) and business opportunities for shopkeepers, but there is also use of scarce water (for hotels), use of wood for heating, additional strains on sanitation, additional congestion in narrow streets, influx of strangers impacting on local relationships and values (for example, sometimes increasing a propensity to purdah for women). The differential local ability to access gains from tourism can add to inequality and threaten the socio-cultural integrity of 'community', and thus threaten other forms of collective action not directly attached to tourism. There is clearly an overall question about the 'ownership' of tourism under these differential access conditions, with a potential to generate conflict over the management of traditional domains of collective action such as water, forest, livestock and the sustained conservation of wildlife (Wood, 1996). However, apart from a few pockets in central Hunza and Skardu-Baltistan, many of these issues are more warnings for the future rather than realities of the present.

Environmental degradation

Nevertheless, these cautions about tourism do interact with other, more inherent, trends in environmental degradation: deforestation; soil exhaustion and erosion; pressure on winter grazing; overcrowding of fertile spaces (that is, alluvial fans); rapid development of growth pole centres, representing pressures on sanitation, drinking water, 'urban' fuel demands, and waste disposal.

Of these, perhaps deforestation is the most significant. While it is difficult to obtain consistent measures of deterioration, there is a variety of evidence to hand from: recall; observation of cut-stumps; inner forest clearance to create new agricultural plots; transportation (legal and illegal) of wood through checkposts; soil erosion; expansion of demand; relative prices and elasticity of demand for alternative fuel sources (paraffin, gas, micro-hydel – though rarely for heating). Ali Gohar, a key forester for the Gilgit–Hunza region reports shrinkage of natural forest in three separate locations after detailed study: Chilas–Darail, Naltar, and Chalt–Chaprote (Gohar, 2002). Each of these forests is under contrasting institutional management, though each contains much ambiguity about rights and applicable legislative frameworks. In the lower Gilgit region of Chilas–Darail, the Pashtun Khans of Swat (to the west) gained ownership rights over the area and have consequently been able to sell large quantities of timber in the market with little regard for the formal constraints of regulation. The state has a precarious presence in the local society, and its local officials are so poorly and unreliably remunerated that a little local bribery goes a long way. In Naltar, with a military high-altitude training interest, the state formally manages but cannot in detail regulate the 'encroachment' of local stakeholders who claim both need and longer-term community-based rights, allocated to them prior to 1947 and the rebellion against Kashmir. These stakeholders are both immediately local at the altitude of the forest as well as from the lower settlement of Nomal, in the main valley 16 miles away. The Chalt–Chaprote forest, at the intersection between the Nagar and Hunza valleys, comprises claims by two communities, though the forest is actually located in the upper, Chaprote, community. It is community-run through rules of allocation for domestic use (heating and building), with regulatory checkposts. But periodically, severe conflict over both the rules and practices of regulation breaks out.

Soil exhaustion is a feature of increasing cropping intensities and also introducing more demanding and fertilizer dependent crops, deriving from population growth, multiple inheritance under conditions of weak alternative income sources, land fragmentation and the need to increase land productivity, and commercial opportunities – as discussed above. Much of the behaviour leading to soil exhaustion can be understood as individual, deriving from family survival imperatives. Nevertheless there are public consequences in terms of: an increased propensity for flash floods to remove fragile top soils from neighbouring plots; and perversely, increased demands for irrigation water at other times of the year. Such treatment of these soils leads to a rapid deterioration of bio-mass, less water retention, and a threat to stable terracing with negative downstream effects. Soil erosion, on the other hand, is closely connected to deforestation as mountain slopes become increasingly denuded of cover, and fragile top soils are irreversibly removed through snow and glacial melt, increasing the propensity to flash floods thus causing further problems lower down in the populated alluvial fans.

With such a livestock dependent society, the management of livestock and its grazing is key feature of collective action. Indeed, it would almost seem that the metaphor used by Hardin (1968) for his Tragedy of the Commons ceases to be a metaphor in this context! The traditional system has been to graze animals in a controlled manner with stored fodder and local common spaces in and around the village and the limited available cover on cultivable land during the winter, and move them up to high pasture during the summer months when the village lands are reserved for cropping, including fodder crops. This system relies heavily upon collective management, including, crucially, the management of conflict, since improper grazing around the village (inadvertent or deliberate) is the most common source of conflict in a society where offence and honour are sensitive issues. Since animals are usually tended by children or young women, managing transgressions has added complications of plaintiff adults disputing with the parents of young herders, in a context of highly imperfect information. However, this traditional system is under threat in a variety of ways. First, where double cropping has been introduced, the grazing 'season' in the village is reduced. Second, this increases either the need for more fodder to be grown (a perverse effect) or for the high pasture season to be extended. Third, this can reduce the carrying capacity of the high pasture even with stable herd sizes. However, fourth, even though there is evidence that herd size per household is declining, the number of households is increasing (though smaller in size as they become nucleated), so that any decrease in numbers of livestock per human capita is offset by an overall absolute maintenance and maybe increase in herd numbers dependent on high pasture. Fifth, add to this the fluctuating influx of large herds from Afghanistan brought over passes by desperate pastoralists and traders who have less knowledge for local rules (and maybe less respect) and where a high proportion of these herds are transient and quickly *en route* to downcountry markets in Pakistan, then the demands for local pasture are especially unstable and uncertain in particular locations.

An aspect of the grazing problem, and its related impacts on soil erosion on mountain slopes and high pastures, is the intensification of settlement and use of the fertile alluvial fans. With population growth and the limited, uncertain prospects to be derived from migration, retention of family land in the scarce fertile areas is an imperative. Multiple inheritance is fragmenting the size of holdings into smaller units per household, and causing it to be cultivated more intensively where ecologically possible. One effect is soil exhaustion as noted above. However, there is also a competition between residential and cultivation space, where it can be observed that overcrowding is encouraging the construction of new residences on erstwhile cultivable land. There is also infilling of common space in the residential locations of the village with negative implications for: casual 'around the house' grazing; sanitation and waste disposal; pollution from domestic smoke; constriction of common social space; mobility of women in traditional households; access for supplies; demand for services (drinking water, electricity, and so on). This intensification of settlement results in additional challenges for collective management as mohallahs alter their profiles of issues to be resolved either in the sense of conflict management or provision of common services. Can the traditional institutions associated with the mosque and the clan cope?

A further aspect of the general overcrowding issue concerns the emergence in some locations of growth-pole centres. These are typically developing at conjunctures of the lower ends of valleys. Sometimes several valleys converge at their lower ends, perhaps feeding into a large valley and river. These geographical points are quite likely to coincide with road routes, since these often 'track' river banks. Some of these locations are growing 'naturally' as growth-poles, since they represent sites of trade, services and other economic activity to which migrants from the surrounding upper levels villages are attracted. Both these and some other locations are 'assisted' in their growth-pole development by decentralization policies of successive governments, building *tehsil* (sub-district) and union (sub-*tehsil*) level headquarters. This has created new opportunities, especially in the construction industry, as officials need to be housed, but also in services such as general retail, house maintenance, fuel supplies, personal service and so on. Again, with the arrival of significant numbers of 'strangers' can the traditional institutions of collective management cope with these new conditions?

Managing the local environment: AKRSP and AKDN

To 'development experts' worldwide, this area of Northern Pakistan has come to attention through the sometimes pioneering work of the Aga Khan Rural Support Programme (AKRSP) and its 'parent' network of institutions in the Aga Khan Development Network (AKDN), which includes other specialist sector organizations in education, health, cultural conservation, sanitation and construction. Although education and health sectors have been established from the 1950s (AKES and AKHS respectively), from the early 1980s, the AKRSP has deliberately set out to address many of the institutional questions raised above in the management of natural and other newly created resources. With a deliberate programme of creating new institutions at the village level, it has sometimes created the medium through which local communities interact with the other AK programmes. It should be noted that the AK (Aga Khan) institutions have been so prominent in the area (especially Gilgit–Hunza and Baltistan) precisely because of its special political status in respect of the central Pakistan government as an area whose ultimate political status remains to be resolved alongside the resolution of the Kashmir question. Thus, while the Pakistan government has a presence in the area through the army and local councils, it has been wary of full-scale political incorporation as a signal to the Indian government that this is in effect Pakistani territory with the implication that across the line of control is therefore Indian territory. This can never be conceded without a full plebiscite on both sides of the line of control with a range of options offered. Under these conditions, the AKDN, with AKRSP in the forefront, has been the development catalyst in the region.

All these AK institutions operate under the general leadership of the Aga Khan Foundation worldwide. Although it has a 'development' interest in areas of significant Ismaili presence, it seeks to work, if allowed, with all communities. Thus AKRSP works among the Sunni communities in lower Chitral, and the Shia communities of Baltistan as well as the Ismaili and Sunni–Shia pockets in the remaining parts of the NAC region. In all of these areas, the erstwhile 'feudal' structures were formally abolished in the mid-seventies, but were not obviously and immediately replaced by other forms of government, as explained above. Thus the rationale for AKRSP can be described in the following terms:

- *Points of departure*: Poverty; institutional vacuum; precarious natural resources under threat.
- *Objectives*: Improved and sustainable productivity of the natural resource base.
- *Rural development strategy*: Enhancement of physical capital via grant and subsidy; levering social capital in the form of village organizations and women's organizations; supported by human capital investment and financial capital accumulation.

The impact of AKRSP on the region has been analysed by many insiders and outsiders, including a series of World Bank evaluations, the most recent of which was launched in February 2002. However the issue for this paper is collective action in environment management – the interface between the social and natural worlds. With this focus, and recognizing that AKRSP has also been operating in a changing socio-economic context beyond its own influence, the analysis of collective action has to reflect a combination of historical and contemporary dimensions.

First, the assumptions made by AKRSP about an 'institutional vacuum' after the abolition of feudal authority structures were too strong and overemphasized. It is clear that many traditional local institutions have continued to perform essential collective management functions in the village. Some of these institutions are religious, performing social functions; some are hierarchical; some are reciprocal. Thus the village organizations (VOs) and women's organizations (WOs) formed by AKRSP are part of pre-existing social structures and as such are deeply embedded into the village institutional landscape, consisting of kin relations, clan structures and power configurations. These relationships affect the character of 'participation' and 'democracy' in these AKRSP organizations. What appears as full-scale collective action is usually reflecting inequalities in influence and power, with the stronger families dominating. While the basis of that domination may have altered, say from being the appointee of the feudal authority towards being the larger landholder and/or receiver of the larger remittances leading to visible positional goods (housing, 4 × 4s, satellite TVs, and so on), the erstwhile stronger families have mainly been the ones to capture the new opportunities.

Secondly, as has been noted, there was and is considerable micro-diversity across AKRSP's area of operation in terms of: location (growth-poles and peripheries); cropping zone; accessibility by road; religious sect; language group; traditional institutions; education levels; migration patterns (downcountry education, employment); availability of remittance incomes; extent of reliance on farming and other natural resource for livelihoods survival; penetration of external markets; commercial production and trading opportunities. The social significance of this diversity for collective action is that families resemble each other less and less. They have less in common to manage and therefore declining interest in maintaining the integrity of membership, rules, adjudication and sanctions. However, although there is a trend shift away from collective social and cultural forms to individualization derived from an increasing nucleation of families partly induced by more family specific and diversified livelihood portfolios, in management terms, this trend is not universal across all sectors. This is where the subtlety of analysis is required, especially where a continued reliance upon some local natural resources is required for all families, partly as an insurance against disaster in the downcountry economy. Thus collective forms of management are retained in irrigation and grazing, even if 'management' now only involves respect for rules and the hiring of other labour to perform the family labour obligations. We will also see below situations where it reappears. And spatially, individualization is more associated with growth-pole centres with collective principles remaining stronger in the more peripheral, remoter villages.

Thirdly, in a more theoretical sense these trends towards individualization have profound implications for the cohesiveness of the shared sense of 'community' and the prospects for common property management. The whole notion of membership of the local community and common property management group is challenged by the socio-economic differentiation between families, with family members developing attachments and loyalties to other institutions, especially through employment but also sometimes through political party allegiances. Back in the local communities of NAC, these crosscutting and often external ties introduce problems of free-riding and the breakdown of collective economic security as the primary identity. Those who out-migrate, even seasonally, reduce their reliance upon these local, collectively managed resources, but cannot actually detach themselves completely due to insecurities elsewhere. Thus they cannot be complete deserters, while finding it difficult simultaneously to be complete members. Their participation is restricted, so that free-riding is more structural and less voluntary, motivated more by competing loyalties to other parts of their livelihoods portfolio than by cheating and the search for unfair advantage, which so characterize the literature on free-riding (for example, Ostrom, 1990 and Ostrom et al., 1994). These trends challenge and set limits to the validity of indigenous, traditional institutions of collective management (that is, clan, mohallah, mosque/jamatkhana/imambarga) as well as the VOs and WOs formed by AKRSP as the preferred institutions for natural resource management (NRM), infrastructural investment and maintenance, and financial transactions within the 'communities'.

Fourthly, as discussed above, the increasing integration of local communities with wider markets and downcountry employment opportunities is responsible for more socio-economic differentiation (that is, inequality) in the villages (Wood, 1996). This has introduced a potential irony, with AKRSP tempted to focus more upon poverty eradication through socially targeting its interventions rather than working with the 'whole community'. This notion of targeting is a new departure for AKRSP and has been hotly debated within the organization and its constituencies, since the whole previous philosophy of its approach has been to underplay internal differentiation at the local level in order to promote community level participation and social capital through institution building. However, a more targeted strategy of poverty reduction, partially in response to demands from donors but also a response to the changing profile of poverty, runs the risk of further undermining collective management principles by acknowledging the limitations of the non-poor families to respond to the needs of the poorest in the community. However, again, the picture is not so simple. We have found that poverty targeting can also bring collective institutions of moral responsibility further into play with local safety net and welfare transfer functions being performed by mosque committees or other such mohallah level organizations, thereby reinforcing or even rejuvenating collective action and traditional charitable instincts and practices (Wood, 2002).

Social capital and non-linear change

To see the both the potential challenges to collective action as well as conditions under which it can be sustained and even enhanced, we can examine four key questions: the continuing reliance upon NRM; the energy problem; the significance of socio-economic differentiation and connection to wider features of globalization; and forms of re-localization through heightened sectarian identity.

Reliance on NRM (public goods, personal resources)

NRM conceived as the management of forest (natural or planted), grazing areas (high or low), wildlife and fish has both 'public goods' and 'personal resource' properties. Motivations for action vary accordingly. The question is: which actions are consistent with sustainability? There is also the problem of whether local conceptions of common property are in fact challenged by other claimants, thus converting the resource to open access, or common pool status. Without an effective local state, the defining and policing of common property claims can only rely upon a common sense of rights and therefore legitimate exclusion, or upon meaningful threats of sanction to produce the same outcome. These outcomes are also affected by time-preference behaviour, which can vary both according to the resource and its public/private property status. Thus forests and pastures, as quintessentially public goods with weak property status and strong open access characteristics, induce high discount rates among users. Indeed the fear of competitive overuse by others can induce depletion behaviour beyond immediate use values or even realistic exchange values. However public goods and personal resources need not always be in conflict. Here we return to the issue of dependency and constraints to resource-hopping. Using the forest depletion comparison between Chilas–Darail and Chalt–Chaprote (see above, and Gohar, 2002). The opportunities for resource-hopping are greater in Chilas–Darail with irreversible depletion already evident due to proximity to downcountry markets and the mobility of forest owners to capture entirely different resources derived from forest profits. This contrasts strongly with the Chalt–Chaprote case in the upper, remoter valleys where resource-hopping is constrained by lack of alternatives, so that personal use-value interests, low discount rates and therefore sustainable public goods management have a greater chance to coincide. There are outbreaks of conflicts, but the collective action institutions can and are called into play for their resolution. In Ostrom's terms (1990 and later publications by her), the sense of the multi-period, self-contract-enforcement game is sufficiently shared as to constrain unworthy behaviour in the present, or at least to permit iterative negotiation in a game sequence. In effect this consciousness becomes our understanding of community and collective action in such a context. By contrast, under resource-hopping opportunities, high discount rates are rational along, therefore, with free-riding, a weakened sense of common property as open access conditions take over, and low investment in either renewing the resource or devising mechanisms of protection. It is not only that collective action fails to prevent this sequence of behaviour with its depletion outcome, but that any ingredients of collective action are positively undermined in the process.

Energy: constraints and opportunities

It is customary to spread gloom on this sector by referring to exponential demand in the face of limited supply leading to: forest depletion; higher costs of fuel purchases; and the inadequacies of the state in providing appropriate large-scale investment and efficient unit costs of supply. While these observations are largely true as contributing to the breakdown of common property forms of management, we can note a significant non-linear development in which social capital is reaffirmed and strengthened rather then giving way to the pressures of differentiation and individualization. The case is micro-hydels. These are small-scale turbines operating as a 'run of river' technology, with some reorganization of the upper flow, concentrating it into channels to create a powerful head before sending it shooting down pipes to the turbine. These micro-hydels mainly produce lighting, but can also power small motors, especially for milling. Generally, the capacity in relation to demand for lighting prevents any use for heating. However the significance of electric lighting cannot be underestimated and the local demand is very high, though as Lawson-McDowall (2000) has demonstrated, not inelastic. The collective action required to manage these units sustainably is considerable through different phases of construction, operation and maintenance and cost recovery. If external grant support is required, this can only be obtained with evidence of a partially matching fund raised in the community to cover subsequent maintenance costs. Thus community level funds have to be raised even before anyone has had sight of the installation. Although a grant might support equipment costs and some technical labour inputs, usually the community has to commit unskilled labour. Again this has to be mobilized, keeping various dimensions of equity between households in mind. A management committee has to be created with responsibility not only for mobilizing labour but also for raising and sustaining the necessary finance. With the micro-hydel installed, it has to be operated in a sustainable manner. This is where long-term collective action is required for everyone's success. A charging system has to be devised which maximizes the take up of available capacity, while avoiding free-riding and the high transaction costs associated with fee avoidance. Over-inclusiveness may only be achieved at a price which is too low to maintain the operation. Exclusiveness at a higher price may be divisive within the community and undermine other forms of parallel collective action (such as forest or grazing management). Exclusiveness combined with means-tested philanthropy can entail high transactions costs, but they can achieve legitimate outcomes in which 'free-riding' by the poor is institutionalized as deliberate policy. Lawson-McDowall (2000) certainly found a higher incidence of such behaviour in Chitral than in his comparative cases in Nepal. With hundreds of these micro-hydels now installed in the region, especially in Chitral, there is clear evidence of successful and sustained collective management. This is occurring even in situations where other common property resource management had broken down, and in some instances has played a role in rehabilitating other, parallel forms of common property management as a bonus outcome.

The overall conclusion here is that while there may be linear trends in the decline of social capital around the management of key renewable natural resources, there can be reversals of such trends if new, highly desired resources are created (such as micro-hydel-sourced electricity) under conditions where resource-hopping to other forms of energy is effectively denied. Coal deposits have been found but not yet successfully exploited, and there would be many other environmental management problems attached to such an initiative.[1] Oil and gas is too expensive to import into the area on a significant scale, even for the more well-off families, though some families are using gas cylinders for domestic heating. Solar panel technology has not yet reduced its unit costs to make it commercially viable in relation to local effective demand. Larger-scale, state-run turbines remain risky as a public sector venture in the eyes of local people due to perceptions of corruption and incompetence.

New patterns of socio-economic differentiation and inequality

Fundamentally, the emerging evidence of poor households scattered among the non-poor, under the reported conditions of socio-economic change due to higher interaction with external markets, confronts locally cherished concepts of equality, homogeneity and mutual interest around common objectives that will produce equally distributed benefits. This challenges also the basis of common cash and labour inputs into joint projects, when returns to households are differential in impact. For example, the erstwhile principle of 'rough equity' in irrigation management (see Wood, 1999 for a comparable point about Bangladesh) is difficult to sustain if some households are acquiring land at the expense of others, and/or cropping it more intensively with high value commercial crops such as potatoes while contributing equal labour input shares as other poorer households.

While not suggesting that we are witnessing classic 'class relations' as experienced downcountry and elsewhere in riverine South Asia, the evidence of increasing socio-economic differentiation and inequalities in

[1] Many market and environmental pollution questions remain: pricing; subsidies, leakage and exports; correlative infrastructure; transportation; geographical and social distribution of benefits and costs; in-migration of specialist labour; waste disposal and water pollution; growth pole attractions for other industries, including other, more problematic (in environmental terms) mining. All these questions relate to issues of ownership over the process both in an early asset sense as well as entailed social and cultural changes towards individualization, commercialization and high discount rates as a deliberate strategy of outmigration.

economic and social measures has to be considered in relational terms as well as distributional ones. That is to say, does inequality also represent exploitation, predation, and individualistic (that is, household) morality or 'amoral familism' (Banfield, 1958)?

Whereas there has been an earlier image of broadly equal peasants as 'subjects' of various feudal arrangements as representing the only form of inequality (lords and tenants), there are now 'surplus value' relationships between educated landowners, tenants and 'landless' labour. Such relations are overlaid by power exercised by strong clans and nuclear families, especially in the deployment of social labour or 'head' levies on 'common interest' projects. In other words, what sometimes appears as fully participatory, collective action is more likely to reflect inequalities of power between clans, extended/joint households and the nuclear elements within them. Meetings are attended, votes cast, funds raised, labour promised – but all under the implicit (2nd dimension – Lukes, 1974) power of the influential local-level leaders. Furthermore, there is an interaction between distributional and relational inequality. The former frequently translates into the latter, as successful migrants and market entrepreneurs convert resources and personal capabilities gained in other arenas to exercise relational power locally. A typical example will be returning army non-commissioned officers who have risen up the ranks after many years of service and have accumulated pension entitlements, which constitute a fortune by local comparison. Other examples are educated sons of the past who have entered government or even better paid (though less corrupt) non-government service, thus gaining incomes, wealth and perhaps even more importantly connections and networks which they are able to deploy as new patrons locally.

At the same time, distributional inequality is also compatible with greater individualization, in which the families moving ahead economically are in effect resource-hopping and weakening their dependence upon local natural renewables. In this way, perhaps reinforced through lengthy absences either in the urban centres of the region itself (for government/army or NGO work) or further away downcountry, they are participating less in the village level institutions even though remnants of their families may continue to reside permanently in those settings. But to the extent that those family remnants are increasingly relying upon remittances and perhaps periods of stay in migrants relatives' houses elsewhere (for example, for education, health or elderly care), their own reliance upon local NRM is reduced. There is clearly a strong gender dimension to these processes, with males typically migrating more than females, with females perhaps joining husbands and other relatives later if the male has successfully relocated. Families locally who are only represented by females are clearly at a disadvantage in institutions of collective action, where the culture endorses males in the public sphere and excludes women. However, with extended families it is still common for a deserted or temporarily alone female (with husband and/or sons outmigrated) to be represented by another male relative. However, the female remains disadvantaged in never being sure that her (or her family) interests are being genuinely represented.

Sectarian identities

Common property management in the region occurs at many different levels of scale in the region. Clearly, forests stretch across different communities and are affected therefore by competing rights and claims to use. High pastures have similar characteristics, despite long-established customary allocation which is perpetually in dispute. Likewise water, with irrigation channels extending from glacial sources over long distances to their final, tail-end destinations with numerous distributories to the intervening communities *en route*. The micro-hydel technologies are adding a new dimension to potential conflict if opportunities are denied tail-enders by upstream demand. Thus 'resource' space may not coincide with community or 'social' space in which multi-layered relationships alongside multi-period games contribute, as social capital, towards prospects for collective action. This lack of coincidence between resource and social space, when the resources are distributed over a wider scale, can be exacerbated by sectarian identities in the Northern Areas and Chitral (NAC) regions of Pakistan.

The religious aspects of settlement patterns reveal, in sectarian identity terms, some 'mixed' villages, certainly mixed valleys, but also subregions dominated by one sect or another. In Chitral, the principal tensions are between Sunni and Ismaili (the latter sometimes are not even accepted as Muslims by the extremist interpreters of Islam among the former). For Ghizer, Gilgit–Hunza, the tensions can be variously between Sunni and Ismaili, and Sunni and Shia, with only insignificant differences as a basis for resource competition between Shia and Ismaili. In Baltistan, the sect identity issue is virtually non-existent with an overwhelming Shia population. There is little doubt that, in mixed areas, sectarian tension is rising, with various 'siege' perspectives intensifying identity and solidarities. This is particularly evident for Chitral, broadly between the Ismaili north and Sunni south, though with some key Sunni dominated valleys in the north as well. Whereas before (that is, up to a decade ago), there could be inter-marriage producing 'clans across sects', thus erecting cross-cutting ties which function to offset exclusive identity, sect and clan is now more clearly differentiated as a basis for social identity thus increasing the likelihood of social closure. At village level, where a mix of sects is present, the sub-geographical unit of mohallah will usually be identified with a particular sect and thus constitute the basis of collective action for resources which can be managed at that micro-level. Thus the institutions of solidarity and identity around the mosque (Sunni), jamatkhana (Ismaili) and imambarga (Shia) may be intensified and become more functional at a localized level. The problem arises when co-operation is required at higher levels of scale, between differently attached mohallahs.

It is important to appreciate that the intensity of these solidarities are not stable. They can lie dormant for long periods of time, enabling inter-sect co-operation to the point where it is simply not an issue. There are numerous examples of mixed Village Organizations (created by AKRSP), with members of different sects sitting together and acting together. However, in the region, the intensity of these solidarities can be affected by events outside the region altogether. Different sects have different religious calendars, so particular events such as the annual Shia Muharram, which involves public processions, can create tension.[2] There are periodic attacks between Sunni and Shia downcountry which can also trigger tension. The events of 1988 remain imprinted on contemporary memory when downcountry Sunni activists were encouraged by the then government of Pakistan to 'discipline' the Shia community for its disturbing loyalties to Shia identities externally, especially associated with the Iranian religious revolution during the 1980s. The killing of a Sunni cleric in Chitral in August 1999 was widely interpreted as a sect-motivated attack as the guilty party was Ismaili. This caused widespread tension between the two communities and a strong counter-reaction to the AK institutions as Ismaili in origin. However, a closer examination of the incident revealed that this was in fact a family feud over land, with the Sunni cleric actually an earlier converted Ismaili. However, it played a strong role in undermining trust in the region, which is only slowly being rebuilt. This has undoubtedly reduced scope for collective action in the mixed communities and valleys, as well as between the valleys of different sects.

These exclusive identities are also variable, strong or weak, according to other, often quite minor and micro contingencies. Micro-level disputes

---

[2] Think of Northern Ireland and the annual marching season.

over grazing which might simply be the result of a non-attentive child, or over irrigation channel maintenance (say, upstream in another village) can easily be understood as a sect based affront and take time to restore. However, a landslide, broken road, or a flash flood might equally bring different sects in a village or valley together in emergency action. These identities therefore have variable and unstable impact upon broader possibilities and needs for co-operation and collective action. It is a picture of fission and fusion, of identities fragmenting and re-forming at different times around different issues. Under such conditions there can never be perfect and complete trust. The social capital is always under negotiation and being reconfigured. This fragility certainly affects people's calculations of others' behaviour, introducing a more instrumental, amoral basis to co-operation. This is the social breeding ground for queried membership, for suspicion, anticipatory cheating and free-riding – all based upon the principle that others must be thinking and behaving similarly. But we should not discount the significance of local NR dependency and a corresponding lack of opportunity for resource-hopping, alongside multi-layered relationships and multi-period games as the set of countervailing forces supporting collective action.

Conclusion

The arguments and evidence presented above are essentially about the relationship between public goods management (as the sustained management of renewable natural resources) and private resource interests in which the opportunity for resource-hopping becomes a key variable in understanding family commitments to local forms of collective action. With the Aga Khan Rural Support Programme (AKRSP) as the major post-feudal institutional player in the region, we can remind ourselves that the basis for AKRSP's institutional strategy since 1982 has been the principle of collective action, with deliberate interventions to create or 'lever' it in, where the principle has allegedly disappeared or gone underground (the 'institutional vacuum'). It is interesting to reflect how far collective action in pre-abolition days was itself induced by strong 'feudal' leadership (the oriental despotism of hydraulic societies, described by Marx as the Asiatic mode of production), thus justifying a replacement 'external enforcer' in the form of AKRSP and its conditional grants for productive physical infrastructure (PPI).

Perhaps the historical judgement about the value of 'feudal' institutions to environmental management has to be balanced by the evidence of community level, sometimes religious centred, institutions for localized water management, rationing forest products, grazing supervision and mutually exchanging labour services during the feudal period itself and not completely lost thereafter.

However, even if there was no institutional vacuum as such, clearly threats to collective action have been identified. Although there is a proper empirical reluctance to accept a simple trend away from collective action, citing, for example, the evidence of resumed collective management around recently introduced micro-hydels, threats to collective action have to be acknowledged. Even with micro-hydels, the ownership group may be quite passive in management terms, in effect franchising out the operation (including distribution and fee collection) and maintenance of the installation to a specialist sub-group. Nothing wrong with that if it works institutionally. The point is that full-scale collective action, involving participation and democracy, has high transaction costs, and very high opportunity costs for members who need to be away from the community for employment and business. The differentiation noted above has to include the diversification of people's income portfolios: villagers are no longer peasant clones of each other.

The increasing uniqueness of the household in these economic and therefore social terms entails individualization, and more household-centred calculations of advantage and disadvantage. These socio-economic processes induce cultural shifts in moral commitments and increase the propensity for free-riding and therefore compliance costs for remaining members (who could be the poorest with their higher local NR dependency and reduced opportunity for resource-hopping).

There are obvious implications of these threats for social capital spreading across from natural resource management to the organization locally of other development activity: infrastructure creation, operation and maintenance; enterprise stimulation supported by credit and savings initiatives; and social protection for either the chronically or transitional poor. There may be sectoral and geographical variations in the significance of these threats, with some cyclical reversals (as in micro-hydels), but the centrality of these threats to the core of any organizational strategy for sustained development in NAC cannot be underestimated. The past is very unlikely to be the future institutionally with basic norms and values on the move.

At the same time, we should also acknowledge local people's agency as expressed through the emergence of many organizational forms, which differ from the original conceptions of the deliberately created village and women's organizations (VOs and WOs) as multi-purpose organizations, albeit levered into creation by the prospect of generous grants for productive infrastructure. The more recent emergence of the *single-purpose* VO, for example, seems to be an example of members perceiving the difficulties of sustaining collective action over a number of parallel objectives, involving complex, interlocked transactions and high degrees of participation. Single-purpose VOs recognize the limits to collective action, with members reducing their intra-transactions accordingly. The proliferation of other 'Os'[3] (LDs, NGs, VDs) and interest groups, business groups and so on is also testimony to people wishing to experiment with organizations that work better for them against particular objectives and social conditions.

It is interesting in these organizational developments to see to what extent villagers are acting out principles of subsidiarity: creating institutions at different levels appropriate to the scale of the management issue and the numbers of stakeholders to be reconciled. But it is equally important to acknowledge, therefore, that the principle of subsidiarity in this context means that an increased responsibility for personal livelihoods is devolved to the lowest level: namely the household or the individual. This helps to explain the widely observed trend of increasingly nucleated households, with declining responsibilities among brothers for each other's children. In other words, a weakening in the longer term of the moral basis for collective action.

References

AKRSP (1999) *Farm Income Survey*. (Edited by Safdar Parvaiz.)

Banfield, E. (1958) *The Moral Basis of a Backward Society*. Free Press.

Gohar, A. (2002) *Institutional issues of forest management in the northern areas of Pakistan*. Ph.D. work in progress. University of Bath.

Hardin, G. (1968) 'Tragedy of the commons'. *Science*, 162, 1243–8.

Lawson-McDowall, B. (2000) *Handshakes and smiles: the role of social and symbolic resources in the management of a new common property*. Ph.D. Thesis. University of Bath.

Long, N. (1994) *Globalisation and localisation: new challenges for rural research*. Paper prepared for the International Seminar on New Rural Processes in Mexico, May 1994.

Lukes, S. (1974) *Power: A Radical View*. London: Macmillan.

[3] At the village and valley level, various new organizations are being created often by local leaders, especially those returning ex-army non-commissioned officers. Thus we see Local Development Organizations, Non-Governmental Organizations (but on a very small, localized scale), and Village Development Organizations which are typically an amalgam of VOs, especially if they split into clan identities soon after the initial grant for productive infrastructure was obtained.

Ostrom, E. (1990) *Governing the Commons: The Evolution of Institutions for Collective Action*. Cambridge University Press.

Ostrom, E., Gardner, R. & Walker, J. (1994) *Rules, Games and Common-Pool Resources*. Ann Arbor, University of Michigan.

Sharma, U. (1980) *Women, Work and Property in North West India*. London: Tavistock Publications.

Streefland, P., Khan, S. & Van Lieshout, O. (1995) *A Contextual Study of the Northern Areas and Chitral*. AKRSP, May 1995.

Wood, G. (1996) *Avoiding the totem and developing the art in rural development*. Prepared for AKRSP Strategy Review, April 1996.

Wood, G. (1999) 'From farms to services: agricultural reformation in Bangladesh'. In Rogaly, Harris-White & Bose (eds), *Sonar Bangla? Agricultural Growth and Agrarian Change in West Bengal and Bangladesh*, pp. 303–28. Sage.

Wood, G. (2002) *AKRSP poverty policy for 2003–8 phase: issues, strategies and dilemmas*. Paper prepared as part of AKRSP's Poverty Workshop series 9.3.02.

Index